From Butterflies to Thunderbolts

Discovering Science with Books Kids Love

Anthony D. Fredericks

Illustrated by Rebecca N. Fredericks

fulcrum resources

Golden, Colorado

For Vicky and Leo Lynott—
whose love and strength were given when they were needed most ...
and whose friendship and support is forever cherished.

Copyright © 1997 Anthony D. Fredericks
Interior illustrations © 1997 Rebecca N. Fredericks
Book design by Bill Spahr

Library of Congress Cataloging-in-Publication Data

Fredericks, Anthony D.
    From butterflies to thunderbolts : discovering science with books kids love / Anthony D. Fredericks ; illustrated by Rebecca N. Fredericks.
       p.   cm.
    Includes bibliographical references and index.
    ISBN 1-55591-946-4 (paperback)
    1. Science—Study and teaching. 2. Science—Study and teaching—Activity programs. I. Fredericks, Rebecca N.   II. Title.
Q181.F8347   1997
372.3'5044—dc21
                                    97-6921
                                      CIP

Printed in the United States of America
0 9 8 7 6 5 4 3 2

Fulcrum Publishing
16100 Table Mountain Parkway, Suite 300
Golden, Colorado 80403
(800) 992-2908 • (303) 277-1623
www.fulcrum-resources.com

Contents

Preface

Many of us may remember science as a subject in which we had to memorize lots of facts, equations, formulas, and miscellaneous bits of information. We often thought of scientists as men in white lab coats with a dozen pens hanging out of the front pocket, oversized glasses, and frizzy hair. Indeed, the typical "mad scientist" that haunts late-night TV movies is portrayed as just that sort of individual.

For many of us, too, our science experiences in the classroom were less than ideal. Most of our science instruction may have consisted of reading textbooks and memorizing long lists of information. We often did not see any relevance between the material we were learning and the "real world." What resulted was a less than enthusiastic response to science and a disdain for anything remotely scientific. For many of us, we were undoubtedly "scientifically illiterate."

Today, our thoughts about learning science and teaching science are dramatically different. We know that science learning, if it is to be both effective and meaningful, must involve children in numerous "hands-on, minds-on" experiences. By this, I mean that children must be offered multiple opportunities to ask their own questions, pursue their own answers, and investigate corners of their world simply because they are interested and because they are encouraged to participate. In fact, of all the academic areas, science is the one that offers multiple opportunities for children to take an active role in their own learning—whether that learning takes place in the classroom or in the living room.

Indeed, science is one of the most dynamic subjects youngsters will encounter in their lifetimes; thus, it stands to reason that it must be joyous, exciting, and purposeful for them while they are in school. To a great extent, children can teach themselves. Adults can serve a tremendous function by being a catalyst and a facilitator—aiding and assisting children in their learning processes. However, we cannot make a child learn. The processes we employ determine whether or not children learn. The climate in the classroom, the encouragement at home, the opportunities to discover, the models we exemplify, the emphasis we place on instruction also determine how much a child will learn.

This book is designed to help you take an active role in the scientific discoveries of your child or the students with whom you work. It is also designed to take advantage of the natural curiosity of children and the inherent scientist inside each and every young person. Your support and promotion of science throughout the day and throughout the year will have a positive and significant impact on how children learn—now and in the future.

Acknowledgments

For support, encouragement, and guidance in this literary effort I am indebted to the following:

The highest accolades and appreciation deservedly go to my daughter, Rebecca, whose creative and dynamic illustrations have enhanced and enlightened this book immeasurably. May others enjoy her artistic abilities!

To Daniel Forrest-Bank, a project editor/copy editor of superior talents, goes my everlasting appreciation for, once again, transforming my ideas into a coherent and cogent book (notwithstanding the almost impossible deadlines he continues to give me).

To Suzanne Barchers, editor and friend—everything!

To my colleague, Brian Glandon, whose friendship is beyond value, goes a standing ovation (with a plethora of applause) for his good humor, constant support, and unfailing camaraderie.

I am particularly indebted to the many teachers and librarians around the country who have invited me into their schools to share the magic, marvels, and mysteries of good literature and good stories with their students. They are the foundation for this book and the inspiration for a literature-rich curriculum in each and every subject area.

Introduction

Science Learning

Science is an exploration of and an investigation into the unknown. Science is learning more about what we don't know ... filling in gaps in our knowledge base, changing old ideas, modifying concepts, and discovering that we don't necessarily have all the answers just because we know a bunch of facts. In some ways, science is a testament to our own innate ignorance—an ignorance born of a desire to know more about ourselves and our world, not one signifying a complete lack of knowledge.

For children, science can and should become a dynamic and interactive discipline. It should allow children to examine new ideas, play around with concepts and precepts, and discover that there is no such thing as a body of finite knowledge. What does this mean for teachers and parents? It means that children must be given a multitude of opportunities to probe, poke, and peek into the mysteries of the universe—whether that universe is their own backyard or a galaxy far away.

Science should also give children a host of opportunities to think, instead of just memorize. Knowing the parts of a frog or the number of planets in the solar system means very little unless youngsters are provided with opportunities to use that information in some useful way. Indeed, science is more than numbers, charts, and graphs—it is a *venture* and an *adventure* of the mind—constantly learning and relearning new data and new ideas. Providing youngsters with opportunities to pose questions about their world, question basic assumptions, or actively seek solutions to various mysteries places a value on the power of the human mind—particularly the minds of children.

Guidelines for Growing in Science

Children need many opportunities to make sense of their world as well as lay a foundation from which future discoveries can emanate. The following guidelines should be considered as markers from which kids can grow in science:

1. Children need to use science information in a practical and personal way. Science instruction should be geared toward offering youngsters many opportunities to put their knowledge into practice, to see science as a daily human activity, and to increase their appreciation of the world around them.
2. Children must take some responsibility for their own learning. Children need opportunities to make their own choices or select learning opportunities based on their goals and interests. Kids who are given

those choices begin to assume greater control over their personal learning and are more willing to pursue learning for its own sake.

3. Children are naturally curious. Using children's innate curiosity about the world can be a powerful motivator—particularly in science.

4. Children are naturally active. The very nature of science implies an action-oriented and process-oriented approach to learning. By this, I mean that children need to "get their hands dirty" in science—they need to manipulate objects, try out different approaches, spill things, break an occasional dish or test tube, handle substances and animals, look around them, taste various objects, and get involved.

5. Children need to be stimulated in diverse ways. Tasting, hearing, seeing, feeling, and smelling are the avenues through which children learn about their immediate environment as well as environments outside the home. These same senses can be integral elements of science, too—signifying to children that the skills they have previously relied on can be used to foster a better understanding about new areas of discovery and exploration.

6. Children need to be engaged in intellectually stimulating encounters with their world. Science provides children with a host of opportunities to question and think critically about their world.

The principles above support the notion that science education, to be productive, requires a partnership between teachers and parents, the joy of learning, and the children's developing curiosity about their environment. Helping kids appreciate their potential for contributing to, not only their personal knowledge base, but the world around them, can be a beacon for a lifelong appreciation of science.

Science and Literature

Children's literature (also referred to as trade books) can and should be a natural and normal part of children's experiences with science. Literature provides kids with valuable opportunities to extend and expand their knowledge of the world around them as well as discover fascinating information in the life, physical, earth, and space sciences. Trade books also help kids develop a rich appreciation for the scientific concepts, values, and generalizations contained within good books. Trade books underscore the idea that science is much more than a dry accumulation of facts and figures. Instead, youngsters begin to realize that books allow them to explore and investigate their immediate and far-flung environment in an arena that has no limits.

By sharing good books with the young people with whom you work you are helping to promote the natural curiosity and inquisitiveness of children. So, too, are you encouraging and stimulating the following:

1. Literature provides children with an ever-expanding array of information in a welcome and familiar format. Youngsters begin to realize that science is not relegated to the pages of a textbook, but can be found on the shelves of the local bookstore or public library.

2. Trade books extend and expand specific scientific concepts beyond information typically presented in textbooks. Literature allows children to explore a topic in greater depth and develop a greater appreciation for all its nuances.

3. Literature relates to children's lives in diverse and divergent ways. A well-written trade book offers readers a variety of information from several angles or several points of view. Youngsters learn that science knowledge is never static—it's always growing and changing.

4. Literature, both fiction and nonfiction, helps children understand the many ways in which scientific knowledge can be shared, discussed, and evaluated. A variety of literature helps youngsters comprehend science as a dynamic subject.

5. Science literature helps youngsters develop positive attitudes toward science in general as well as toward their role as scientists.

6. Current literature provides children with new information and knowledge unobtainable in any other format. Topics in which new discoveries are being made at a rapid rate (e.g., space exploration, medical research, tectonic plate theory) can be shared through recent publications.

7. Science literature stimulates creative thinking and problem-solving abilities in a variety of contexts. These intellectual skills are necessary, not only in a scientific environment, but throughout a person's academic career.

8. Science trade books open up the world and draw young readers in to make their own self-initiated discoveries. In many ways, science literature encourages kids to ask their own questions and provides the impetus to initiate their own investigations. It stimulates their natural inquisitiveness and enhances their appreciation of the known and the unknown.

9. Lastly, learning science through literature is fun. There's no question that kids would rather read a trade book than a textbook. Literature gives them a vehicle with which they can explore, discover, investigate, and examine the world in which they live.

Science literature becomes a powerful motivator for the elementary classroom as well as for the natural activities that parents and children can share at home. You are encouraged to use this book as a stimulus to your child's or student's natural tendency to seek answers to their innumerable questions. You are further encouraged to use the trade books cited within this volume as "instigators" for scientific discoveries and investigations throughout the months and years ahead.

Science as Process

For many years, school science programs were designed to give children lots of information, have them memorize that data, and then ask them to recall the information on tests and quizzes. That type of teaching may be a significant reason for students' less-than-enthusiastic response to science, because it does not allow for the active involvement of students in their own learning, nor does it allow opportunities for students to think critically about what they are learning.

My own experiences as a teacher have taught me that when children, no matter what their abilities or interests, are provided with opportunities to manipulate information in productive ways, learning becomes much more meaningful. I refer to this as a process approach to learning, which provides children with an abundance of projects, activities, and instructional designs that allow them to make decisions and to solve problems. It implies that kids can manipulate, decide, solve, predict, and structure the knowledge of science in a host of meaningful ways. When teachers and parents provide opportunities for children to actively process information, learning becomes much more child-centered and less textbook-based. This results in an exposure to science that is expansive, integrated, and dynamic.

Combining hands-on experiences with inquiry-based science (in which children ask their own questions and pursue the answers to those questions) gives children some control over what they are learning and why they learn it. Thus, children are provided with multiple opportunities to take an active role in science—a role that will be meaningful, exciting, and productive.

The investigation of science is also dynamic! That is, children's daily contacts with the scientific world involve a constant interaction between the known and unknown. New ideas are discovered and others are modified, strengthened, or rejected. What helps children develop a scientific outlook is the processes to which they are exposed in school and at home. A process approach to science stimulates divergent thinking and provides a means for children to investigate their world based on what they know and what they wish to discover.

The following seven processes, which are embedded in the literature and activities of this book, are designed to help you create a stimulating learning environment for kids—whether you are a teacher or parent. Using selected examples of children's literature, these processes are presented in detail. What is important is the need to offer children multiple opportunities to use literature (nonfiction and fiction) as a springboard to new and varied learning opportunities throughout the world of science. In so doing, we can assist children in becoming active and lifelong learners who appreciate the role of science in their everyday lives.

Observing

Observation involves all the primary senses: seeing, hearing, smelling, tasting, and touching. It is how we react to our environment, and it is the source of knowledge humans employ most. Children sometimes tend to overrely on their observational powers or do not use them in concert with other investigative abilities. When children are provided with opportunities to evaluate and question their observational skills, they gain a sense of the importance of this process. Scientific skills are enhanced when children use observation in combination with other processes, such as predicting and experimenting.

**Fredericks, Anthony D. *Clever Camouflagers.*
(MINOCQUA, WI: NORTHWORD PRESS, 1997).**

Observing Activities
1. Take children for a walk around the block, down the street, or through a backyard. Invite them to note each and every animal (big or small) they can. Afterward, ask them to record all the animals they saw. Ask kids to compare their lists.
2. Ask children to try and remember the color(s) of their pets. Have them write those colors on a master chart. Invite children to compare pet colors of various species and types of animals. What similarities or differences do they note?
3. Take children for a walk through the neighborhood. Ask them to record the names of all the animals they see. Upon your return invite children to transfer their data to a large wall chart with the names of animal groups (mammals, birds, fish, amphibians, insects). Ask them to speculate why one type of animal was seen more frequently that another. Are some types of animals more easily seen in nature than others?
4. If possible, obtain a copy of the National Geographic Society video, *The Great Cover-Up: Animal Camouflage* (Educational Services, Washington, D.C. 20036; catalog no. C51311), which alerts children to several different forms of camouflage that animals may use, including deceptive coloring, shape, and behavior.

Questions to Share
1. Were you able to observe more than you expected?
2. Can you offer an explanation for your observations?
3. Have you ever seen or heard anything similar to this?
4. Would your observations be identical if you were to do the same things again?

Classifying
Classifying is the process of assigning basic elements to specific groups. All the items within a particular group share a basic relationship that may or may not be reflected in other groups. As new ideas are encountered they are added to previously formulated groups on the basis of similar elements. Classifying enhances scientific comprehension because it provides children with the opportunity to relate prior knowledge to new concepts.

Pringle, Lawrence. *Dinosaurs, Strange and Wonderful.*
(HONESDALE, PA: BOYDS MILLS, 1995).

Classifying Activities
1. Invite children to cut out magazine pictures that depict certain types of dinosaurs. Ask children to create two or more different collages representing different categories of dinosaurs (large and small; meat-eaters and plant-eaters; land and aquatic, etc.).
2. Invite kids to create mobiles or murals illustrating some of the dinosaurs depicted in the book.
3. Invite children to measure the heights of all the people in the classroom or at home. Have kids divide the class into two or more categories (tall, medium, small; students over 4 feet and students under 4 feet; etc.) and assign individuals to a variety of different groups.
4. Invite children to collect bones from various animals eaten at home (chicken bones, steak bones, fish bones, etc.). Encourage them to create a display of large and small animals based entirely on their bones.

Questions to Share
1. How are these items related?
2. How many different ways could these items be grouped?
3. Are there any similarities between these items and something else you may have seen at home or in your neighborhood?
4. Are there other categories in which some of these items could be placed?

Measuring
Scientists are constantly measuring. Measuring provides scientists with the hard data necessary to confirm hypotheses and make predictions. It yields the firsthand information necessary for all other stages of scientific investigation. Measuring includes gathering data on size, weight, quantity, and number. For obvious reasons it is important that this information be accurate and specific. It is also a valid means of making comparisons using very definite terms, rather than using indefinite language such as large, small, huge, heavy, light, and so on.

Simon, Seymour. *Storms.*
(NEW YORK: MORROW, 1989).

Measuring Activities
1. Invite children to measure the amount of rain that falls in your area over a period of one week, two weeks, three weeks, and four weeks (an inexpensive rain gauge [$4.95; catalog no. 57-180-2426] can be obtained from Delta Education, P.O. Box 950, Hudson, NH 03051).
2. During the next thunderstorm, ask children to judge the distance of a lightning strike by timing how long it takes to hear the thunder (time the interval).

3. Invite children to make charts of storm data that vary from storm to storm (rainfall, duration of storm, wind speed, temperature, etc.). These can be kept in an ongoing journal or diary.
4. If possible, take a field trip to the local TV station and watch a weather broadcast. It may also be possible to interview the station meteorologist and obtain some firsthand information on weather predicting. What are some of the instruments used to measure weather?

Questions to Share
1. Why do you think these two measurements differed?
2. Do you feel more data is needed before we go on?
3. Do you think we should measure these again to see whether we are truly accurate?
4. If the size (weight) of this object was larger (smaller), how do you think it would affect our experiment?

Inferring
Children often need to make conjectures about *current* events. There are two types of inferring: deductive (going from the general to the specific) and inductive (going from the specific to the general). Making inferences requires children to have a sufficient background of personal experiences as well as the opportunities and encouragement to draw tentative conclusions or explanations.

Berger, Melvin. *Germs Make Me Sick.*
(NEW YORK: HARPERCOLLINS, 1995).

Inferring Activities
1. Before reading this book, invite children to speculate on the different ways in which germs can enter the human body. Have children compare their list with the examples given in the book.
2. Read the book to the bottom of page 9. Close the book and ask kids to infer the different ways in which the body keeps out germs.
3. Have kids list several common childhood illnesses or diseases. Have children speculate on the causes for each of those diseases. Is it possible to tell what type of illness a person has just by looking at him or her?

Questions to Share
1. Why do you believe that?
2. Do you have any reason for saying that?
3. Can you think of any other possibilities?
4. Why didn't this come out the way we wanted it to?

Communicating
Communication is the means by which information is shared and disseminated. It involves not only interacting with others but organizing data so that

it can be effectively passed on to others. Communicating can take many forms, including gestures, verbal and written responses, reading, listening, showing, and questioning. An effective communicator is one who is able to organize ideas in such a way that they will be immediately comprehended by others.

Arnold, Caroline. *Giraffe.*
(NEW YORK: MORROW, 1987).

Communicating Activities

1. Invite students to create their own form of sign language and to present a summary of the book to others using their "language." You may want to show them a book on sign language to give them some ideas.
2. Invite children to study pictographs or hieroglyphics and make a summary report of the book.
3. Invite children to choose several sentences from a specific section of the book and to rewrite the sentences in random order. Give them to others to write in the correct order.
4. Invite children to describe a giraffe without using words. What methods of communication are most easily understood?

Questions to Share

1. Why is this easy (hard) to understand?
2. Can you show us another way to communicate this information?
3. What other information about this project do we need to pass on to others?
4. Why is it important for us to write (read, tell) this information?

Predicting

Scientific investigation is a constant process of making predictions. Predicting is the process of extrapolating information based on a minimum amount of data or on information already known. The scientist then tries to confirm or refute the prediction based on the gathering of new data. Predictions provide scientists with a road map by which they can conduct their experiments. They provide goals—albeit tentative ones—but at least something to aim for. The data-gathering process provides scientists with the evidence they need to verify their original predictions.

Kahl, Jonathan. *Weatherwise: Learning about the Weather.*
(MINNEAPOLIS, MN: LERNER PUBLICATIONS, 1992).

Predicting Activities

1. Invite children to make predictions about the next day's weather. List the predictions on a large sheet of paper. On the next day, share the

weather report from the newspaper to determine the accuracy of the predictions.

2. Invite children to predict the high and low temperatures for the day or for the week. Check the data in the local newspaper to confirm the predictions. Ask children to determine how their predictions could be made more accurately.

3. Invite children to list the preliminary information they would need in order to make predictions concerning (a) the speed of wind, (b) the amount of sun, (c) the types of clouds, or (d) the temperature for a particular day of the week.

4. Invite children to look through the newspaper to locate predictions other than the weather (horoscopes, sports scores, etc.). Discuss how these predictions are made.

Questions to Share

1. How did you arrive at your prediction?
2. What makes you feel that your prediction is accurate?
3. What evidence do you think we need to confirm or reject that prediction?
4. Do you have a reason for saying that?

Experimenting

By definition, a true scientist is one who is constantly experimenting. Through experimentation, ideas are proven or disproved and hypotheses are confirmed or denied. Experimentation involves the identification and control of variables in order to arrive at a cause-effect conclusion. Experimentation also involves manipulating data and assessing the results. Children need to understand that they conduct experiments every day, from watching ice cream melt to deciding on what clothes to wear outside. Scientific experimentation, however, involves a more formalized process, albeit one that also touches our everyday activities.

Reigot, Betty. *A Book about Planets and Stars.*
(New York: Scholastic, 1988).

Experimenting Activities

1. Invite children to invent a telescope using a paper towel tubing and a series of concave or convex lenses (available through most science supply houses).

2. Invite kids to create a model of the solar system using the examples in the book. Have them experiment with different materials in the construction of their models.

3. Invite children to invent a new planet. Where should it be placed in relation to the other planets? What would it look like? What would be some of its distinguishing features and characteristics?

4. The excellent video, *Exploring Our Solar System* is available from the National Geographic Society (Educational Services, Washington, D.C. 20036; catalog no. C51356). Before viewing the film ask children to discuss some of the experiments scientists would like to conduct with regard to selected planets in the solar system (e.g., "Is there life on other planets?" or "What is the composition of certain planets?").

Questions to Share

1. What else could we have done to arrive at this conclusion?
2. Is there another experiment we might do to arrive at the same conclusion?
3. Do we need any more evidence before we can say that?
4. Why do you think we did this experiment?

Process Webbing

The value and significance of process skills as necessary elements of science have been emphasized by many educators. Not only should children understand the processes of science, but they must be able to initiate their own self-discoveries and investigations with those skills. In doing so, they will develop an appreciation of their role in science learning and begin to see science as an active and ongoing exploration of the world around them.

By the same token, the use of science trade books offers youngsters opportunities to pursue a topic of interest in greater detail and depth. Trade books also provide innumerable possibilities for utilizing process skills in meaningful and relevant scientific explorations.

The ideas above offer adults many opportunities to share science trade books with youngsters through a variety of interactive and creative activities. Building upon those concepts through a strategy known as "process webbing" is a way teachers and parents can underscore the importance of process activities within a single trade book. This technique effectively blends process skills with a single trade book to open up new avenues of discovery and investigation for children. You will discover the technique adaptable to many different types of nonfiction and fiction books.

Indicated below is a description of a "Process Web" for one book—*Volcanoes* by Seymour Simon. Note the inclusion of both process questions and process activities. Equally noteworthy is the fact that children are engaged in generating these ideas and extensions. Because the kids who will read this book are familiar with a process-oriented science instruction, the generation of relevant ideas is both natural and comfortable. The effectiveness of process webs is because they can be constructed by teachers, parents, *and* children as cooperative learning ventures.

Volcanoes
Seymour Simon
New York: Morrow, 1988

Observing
1. Obtain samples of volcanic ash and/or lava and show them to children. Invite kids to compare the feel of these substances with the descriptions in the book.
2. Invite children to draw their own interpretations of the two gods—Vulcan and Pele. Gather the pictures in a scrapbook.
3. Invite children to compare volcanic ash to the ash from other types of materials (paper, wood, etc.). What is similar? What is different?
4. Invite children to create a flip book illustrating the sequence of activities during a volcanic eruption.
5. Have youngsters watch the filmstrip *Earthquakes and Volcanoes,* available from the National Geographic Society (Educational Services, Washington, D.C. 20036; part of the *Discovering the Powers of Nature* series [catalog no. 03237]). Invite them to generate a list of adjectives (to be recorded in a homemade book) to describe the actions they observed.

Classifying
1. Invite children to place the titles of the four different kinds of volcanoes on separate sheets of paper. Invite them to draw illustrations of selected examples (from around the world) of each type of volcano on the paper.
2. Have children create a scrapbook that classifies volcanic rocks (lava, pumice, etc.) and types of lava (aa, pahoehoe).
3. Invite kids to construct comparative charts of volcanoes according to different climatic regions of the world (e.g., how many active volcanoes are located in tropical regions vs. how many active volcanoes are located in polar regions?).
4. Have children make charts of the dormancy periods of selected volcanoes. For example, which volcanoes have remained dormant the longest? Which volcanoes have had the most recent eruptions? Where are the most dormant volcanoes located? Where are the most active volcanoes located?

Measuring
1. Invite children to locate information on the eruption rates (the length of time from the start of the volcano until it "settles down") for different volcanoes. How can they account for the wide variation in rates?
2. Invite kids to measure the temperatures of different household items (e.g., boiling water, microwave dinner, etc.) and compare those temperatures with the temperature of molten lava (comparative charts can be constructed and posted).

3. Encourage children to investigate the heights of different active and inactive volcanoes around the world. During a volcanic eruption, how much of the mountain is lost?

4. Invite kids to compare the time periods of volcanic eruptions and earthquakes. Why do volcanic eruptions tend to last longer? Why do active volcanoes and major earthquakes occur in the same areas of the world?

5. Have youngsters obtain data on the relative speeds of different types of lava. How fast does quick-moving lava flow in comparison with slow-moving lava?

Inferring

1. Invite students to compare the photographs in this book with volcano photos in other books. What similarities are there? What kinds of differences are noted? How can they account for the differences in photos of the same volcanoes?

2. What happens when you shake up a bottle of soda pop? How is that similar to the action of a volcano? Why?

3. Invite children to develop a pantomime in which they simulate the actions of the volcano, lava, surrounding territory, and so on.

Communicating

1. If possible, divide children into several small groups and invite each group to list as much background information as possible about volcanoes. Have each group chart their data in the form of an outline or semantic web.

2. Have children watch a video of a volcanic eruption (e.g., *The Violent Earth,* available from the National Geographic Society [Educational Services, Washington, D.C. 20036; catalog no. 51234]). Have children pretend they are actually at the site of one of the eruptions. Solicit their reactions during the "eruption."

3. Have youngsters draw pictures of what their geographic area would look like after a volcanic eruption.

4. Invite children to study famous volcanoes in history (Krakatoa, Mount Fuji, Vesuvius, etc.).

5. Invite children to write letters to people who have lived in the area of a volcanic eruption (e.g., the people who lived near Mount St. Helens) and solicit information about their experiences.

6. Ask children to investigate the myths and legends of volcanoes compared with modern scientific knowledge.

7. Invite youngsters to develop a "volcano drill" in the event a volcano were to erupt in close proximity to the school or town. What precautions should be taken? What preparations should be made? How would a "volcano drill" be similar to or different from a fire drill or earthquake drill?

Predicting

1. Invite children to make some predictions on the length of time needed for an area surrounding a volcano to recover from a volcanic eruption.
2. Show one of the "before" photos in the book and have children make predictions on what a volcano will look like after an eruption (they may wish to draw illustrations). Afterward, compare their predictions with a succeeding photo in the book.
3. Invite kids to respond to some of the following questions—"How long will it take for animals and plants to return to an area of a volcanic eruption?"; "Which animals will return first?"; "Which plants will return first?"; "How long will it take to clean up an area?"

Experimenting

1. Take two paraffin blocks and cut them into the shape of earth crust plates. Put them on a hot plate and slowly move them in opposite directions (using heavy-duty gloves) to examine how plates move and react.
2. Boil several eggs until they crack. Invite kids to explain the similarities between the egg shells and the crust of the earth.
3. Obtain some volcanic ash. Mix different amounts with equal amounts of potting soil. Fill several compartments of an egg carton with the different mixtures and plant several vegetable seeds in each compartment. Have youngsters compare the relative growth rates of the vegetables. In which growth medium do the seeds germinate first? Which one is most conducive to healthy growth? How does the amount of volcanic ash affect the germination and growth of plants?

Process webs provide opportunities for children to tie together scientific concepts with relevant and meaningful trade books. Youngsters can begin to understand that the processes of science are not necessarily restricted to the science classroom, but rather that they can be used and applied in many different situations. The linkage of process skills with literature also provides children with opportunities to make science more relevant to other areas of their lives.

A process approach to science supports the natural curiosity of children and enhances their active involvement in the learning process. So too does it allow children to begin their own investigations—investigations based on a solid foundation of support and encouragement.

This book contains a host of activities, projects, and ideas that emphasize the processes of science. The focus is on using children's literature to promote those processes and to assist children in discovering the excitement and thrill of scientific inquiry.

How to Use This Book

This book is designed to offer children opportunities to learn about science through a "hands-on, minds-on" approach to learning coupled with exciting and motivating children's literature. This approach has been shown to offer youngsters unique and stimulating examinations of the world in which they live. So too will children be able to become active learners through experiments, projects, and investigations in all areas of science supported by some of the best children's literature available.

Book Selection

There are thousands of children's books available for youngsters and more than 4,500 new children's books are published each year. Many of those books are excellent additions to the home or school library. The children's literature selected for inclusion in this book includes examples of the "best of the best"—trade books that meet high standards of accuracy, interest, and design. Several important points were considered in deciding on the books featured in these pages:

1. Children's librarians in both public schools and public libraries were consulted on their recommendations of the best literature for promoting science skills and attitudes.
2. Teachers were asked to share their ideas on those books that engender positive responses to science and foster children's enjoyment of the natural world.
3. Award-winning children's literature was also considered in compiling the lists of books included here. Included are Caldecott Medal and Honor books, Newbery Medal and Honor books, Children's Book Award books, American Library Association Notable Books for Children, Reading Rainbow Feature books, Boston Globe/Horn Book Awards, and outstanding books cited in *Book Links* and *Science and Children*.
4. The books included here were selected because they are valuable as standards of high-quality children's literature and are also important in promoting science-related experiences. They represent both fiction and nonfiction examples in a host of scientific disciplines.
5. All of the titles selected for this book extend and enhance children's appreciation of science. Through these books, children can examine and explore the world around them as well as engage in meaningful science-rich experiences with adults.

6. The books represented here are easily accessible in most school or public libraries. Additionally, each of the literature selections (including both the focus books and the related literature for each section) is currently available in any metropolitan bookstore. Ease of accessibility was a primary consideration in the determination of the books presented here.

7. Finally, these books were chosen for their enjoyment as well as their scientific accuracy. The intent is to stimulate an appreciation of science through a participatory approach—one that is grounded in authentic, dynamic, and engaging trade books. Thus, children are able to enjoy science as a natural and normal part of their everyday lives. They also appreciate trade books as vehicles with which they can make those discoveries.

Learning with Literature

Teachers and parents should feel comfortable in helping children begin their own investigations, ask their own questions, and pursue their own discoveries in a host of activities. The ideas in this book provide youngsters with a chance to become scientists and to look at all the dimensions of a favorite topic or a subject of personal interest. Please keep in mind that there is something for everyone in these pages. More than three hundred activities, projects, experiments, investigations, observations, and other creative ideas are sprinkled throughout the book. You and the children you work with are invited to select those pieces of children's literature, those areas of science, or those activities with which you are most comfortable and which encourage kids to initiate and pursue their own questions about the world.

There is no special order to the activities herein. Sit down with a child and flip through the pages. What do you see? What interests you? What books would you like to read? There are plenty of ideas for every child and for every adult who supports a child's learning. Please keep in mind that these activities are only suggestions; you and the children with whom you work are encouraged to modify, adapt, and alter selected activities according to the season of the year, where you live, or your own unique and singular questions about science. Just keep in mind that the best activities are those that support the natural curiosity and inquisitiveness of children and provide opportunities for children to become active and enthusiastic learners.

You and the children don't need a lot of expensive equipment or materials. Many items can be purchased inexpensively or built very cheaply at home. Indeed, some of the best investigations will take place when youngsters actively participate in the gathering of supplies and the construction of selected items such as bird feeders, worm houses, kites, terrariums, and rockets. There are a few activities that require adult supervision, but, for the most part, youngsters should be given a free hand in the design and production of selected projects.

All of the activities are built around the use and sharing of quality children's literature. For each book, the following is included:

1. A bibliographic entry

2. A brief summary of the book

3. Questioning

    For each book there is a selection of questions to share with children. These questions are "open-ended" questions—meaning that there are no right or wrong answers. The intent is to offer youngsters opportunities to discuss what they have read and to stimulate their divergent and creative thinking.

4. Investigating

    Each selected book includes a major book project. This activity is an intensive and stimulating look into a specific area of science and provides a close and exciting examination of an idea, topic, or subject discussed in the pages of the book.

    It is suggested that this activity be introduced to children as a major component in their understanding of the book's concepts as well as an intensive examination of the subject in detail. The construction of terrariums, animal houses, special machines and apparatuses, and selected devices is part of these activities.

    In some instances, adult help or supervision is necessary to satisfactorily complete the suggested task(s). Be sure to encourage youngsters to take an active role in the design and implementation of any related projects. Also, invite kids to elaborate or extend the activities in a variety of ways—according to their interests or self-initiated questions.

5. Additional related activities (one or more of the following)

    (a) Writing: Here children are provided opportunities to inquire about selected topics or to send away for additional information on a particular subject. Included too are projects such as maintaining a diary over an extended period of time, recording field notes in a log book, or corresponding with youngsters in other parts of the country.

    (b) Exploring: Ideas for extending the topic of a book into your local community or neighborhood are offered. Here children will have a chance to see science "in action" and relate what they have learned in a book to the world outside their home or classroom.

    (c) Experimenting: Youngsters are given unique opportunities to examine a topic through a range of carefully designed experiments. As mentioned previously, these experiments do not require expensive equipment or dangerous chemicals. Rather, they offer insights into an area of science so that children will be able to participate freely in the examination of their world.

    (d) Creating: This section provides youngsters with various ways to express themselves artistically through the creation of mobiles, dioramas, posters, displays, and other artistic ventures. It is suggested that these creations be publicly displayed for all to enjoy.

(e) Extending: Teachers know that when science becomes an element in every area of the elementary curriculum, youngsters get more excited about science and begin to understand the relevance of science to the world around them. This section provides varied opportunities in social studies, language arts, music, art, physical education, and math for children to examine and explore their environment.

(f) Cooking: Cooking is a magical experience for many children. In this section youngsters are provided with a host of scientific principles to use in the kitchen to create a range of tasty treats and dishes. Baking brownies that use products obtained from the rainforest *(environment),* creating a vegetarian stew *(plants),* "inventing" a snack low in fat and cholesterol *(general),* and preparing a dinner from the sea *(oceans)* are just a few of the cooking projects kids can investigate.

(g) Special investigating: This section allows children to set up and examine a special component of a selected book. Many of these activities are long-term—those that may last for several weeks or several months. By looking at science over a longer period of time, kids can examine the growth and development of plants and animals and other scientific principles in an unhurried fashion.

(h) Dramatizing: Kids need many opportunities to express themselves in plays, storytelling experiences, and creative expression. These activities allow youngsters unique opportunities to extend their science learning into other areas of their lives and to understand how science is part of everything we do.

(i) Expressing: Dance and music can be wonderful extensions of any science program. Here kids create and perform projects that allow them to understand how the arts can be part of science and how science can be part of the arts.

6. Incredible facts (and accompanying questions related to the book's topic)

   This section introduces children to some of the most amazing and incredible information about a topic. Here children learn fascinating data and unbelievable facts about a selected subject. So too are children given opportunities to examine and discuss that information with a variety of investigative questions.

7. Related literature

   This section offers a listing of extended readings related to the primary book. Most of these books can be found in your school library or local public library and offer youngsters opportunities to pursue an area of science in greater detail.

Children may elect to pursue the major project or choose to participate in one or more of the other ventures. As you assist children in selecting appropriate activities be mindful of their interest and abilities, the available time and materials, and the degree to which you choose to participate.

You and kids should feel free to use the activities for any book in any order you wish. By the same token, feel free to reject some activities, modify others, or add your own. You will discover that there are no limits to science when imagination and creativity combine with good literature to present learning opportunities for every child. Indeed, an active exploration of science can begin with a good book and extend far beyond the print and pictures.

The activities you and kids select within the pages of this book are intended to open their minds to the marvels and mysteries of science and assist them in appreciating the role of science in their everyday lives. These explorations are grounded in children's literature—quality books that provide insights, offer valuable information, and extend learning opportunities. When quality literature is made part of any scientific exploration, children can become involved in activities and gain experiences they would otherwise not be exposed to in any other format, including textbooks.

I sincerely hope you and the children with whom you work will discover a plethora of exciting, stimulating, and magical adventures within the pages of this book. Equally, I hope that those same children will enjoy science as a worthwhile and valuable learning experience that opens their eyes, extends their possibilities, and expands their horizons now and for many years to come.

Part I

Animals—
From A to Zoo

Chapter 1

Bugs

Nancy Winslow Parker
and Joan Richards Wright

NEW YORK: MULBERRY, 1987

Book Summary

A fascinating examination of fifteen common insects, this book is a creative look into the world of these tiny creatures. Rich and colorful illustrations, a mix of fiction and nonfiction, and an engaging text combine to form a book that children will turn to again and again.

Questions to Share

1. Of all the bugs mentioned in the book, which one was your favorite? Why?
2. How many of the bugs described in the book have you seen around your house or neighborhood? Which one frightens you the most?
3. Why do you think some people are afraid of bugs? Why do people try to get rid of bugs from their homes?
4. Are any of the bugs mentioned in the book helpful to humans? How?
5. What is the most unusual fact you learned about bugs from reading this book?

Major Book Project

Insects live in a variety of locations as well as a variety of "homes." Children can collect and keep insects for extended periods of time by constructing the following insect house. Encourage children to keep a record of what insects do in the "house"—how they behave, what they eat, and other daily patterns. These "field notes" can be a valuable guide to many different types of insects in the local area. (*Note:* Much of the construction of this insect "house" should be done by an adult—particularly those steps that require the use of sharp cutting tools.)

1. Provide each child with an empty oatmeal box. Following the illustration on the next page, draw three separate "windows" on the outside of the box using a magic marker.
2. With a sharp knife or craft cutting tool, cut out the "windows" from the box. Remove any remaining label from the outside of the box.

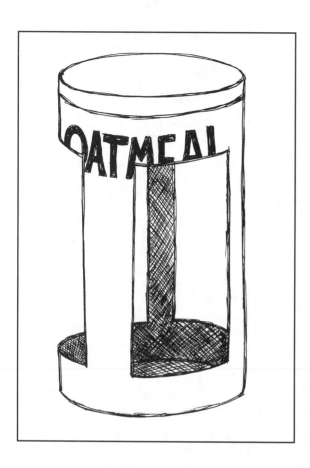

Invite each child to paint the outside of the box with poster paints or magic markers.

3. Stretch an old nylon stocking over the outside of the box (nude or tan colored stockings work best).

4. Invite each child to place one or two twigs, a few leaves, and perhaps some grass clippings into the box. Take children outside to collect two or three bugs to place in the house to observe for approximately two weeks (when this activity is completed, be sure to return the bugs to their original habitat).

5. A soda bottle top can be placed upside down in the bottom of the house and filled with water. Fish flakes (obtainable at your local pet store), oatmeal, cornmeal, or other grains can serve as food for the inhabitants (you may wish to check with your local pet store for other possible foods).

Creative Dramatics

Ask each child to select one of the insects mentioned in the book or another bug of their own choosing. Invite each child to demonstrate the movement of that insect in a designated area. For example, for a mosquito, children can extend their arms and "buzz" around the room; for a centipede, children can wriggle across the room; for a cricket, children can leap around on their hands and knees. Provide opportunities for children to describe their movements and why those movements may be unique to each selected insect.

Across the Curriculum

Math

Invite each child to keep a log book of the numbers of selected bugs located in a specific area (a room in the house, a section of the classroom, a plot of land in the backyard). Encourage children to record numbers of bugs observed during a designated part of each day (from 3:30 to 4:00, for example) over a selected period of time (such as one week). Invite children to create a chart or graph that records those numbers and that can be displayed for others.

Language Arts

After each child has had an opportunity to read the book or listen to an adult read the book aloud, ask children to create a make-believe insect that is yet to be discovered. The make-believe insect should be illustrated, named, and a brief paragraph written telling where the insect would live, what it would eat, if it is helpful or pesty, and so on. A special book about this new insect can be created by stapling several sheets of paper together. Be sure to offer opportunities for children to share their finished work.

Social Studies, Reading

Many insects live in social groups. Ants and termites, mentioned in the book, are two examples of social insects. Share the book *Ant Cities* by Arthur Dorros (New York: Harper and Row, 1987) and discuss with children the similarities between ant cities and the cities that people live in. What are the realted elements of those two types of cities? What elements of one type of city are missing from the other type of city? Why do children think the author titled the book as he did? (*Note:* Directions for the creation of an ant farm are included in the book *Ant Cities*. Children may wish to create their own ant city and observe the social behavior of ants firsthand.)

- 🦌 **The body temperature of butterflies must be at least 81°F before they can fly.**
  - *What is the difference between a butterfly's body temperature and your body temperature?*
  - *How many different types of butterflies can you find near your home?*
  - *What is the world's largest butterfly? The world's smallest?*

- 🦌 **A mosquito, filled with blood, is able to fly carrying a load twice its own weight.**
  - *How much weight can you carry while walking?*
  - *What is your body weight?*
  - *What is the difference between the weight you can carry and your body weight?*

- 🦌 **A queen termite produces eleven million eggs every year for fifteen years.**
  - *How many eggs a month does a queen termite produce?*
  - *How many eggs per day?*
  - *Look around your house or around old logs for termites. How are they similar to the insects in* Bugs?

- 🦌 **Flies have tiny balancing organs at the back of their bodies, which help them land upside down.**
  - *What is the largest fly in the world?*
  - *What is the difference between a horsefly and a housefly?*
  - *How much does the average fly weigh?*

Related Literature

Brinckloe, J. *Fireflies!* (New York: Macmillan, 1985).
This beautiful story captures the freedom of fireflies on a summer evening.

Carle, Eric. *The Grouchy Ladybug.* (New York: HarperCollins, 1985).
A bragging ladybug becomes better behaved as it learns something about getting along with others.

Hornblow, L., and A. Hornblow. *Insects Do the Strangest Things.* (New York: Random House, 1989).
A book full of amazing and delightful facts about some of the most unusual insects in the animal kingdom. A sure delight!

Julivert, Angel. *Ants.* (Hauppauge, NY: Barron's, 1991).
Rich, vibrant, and colorful illustrations highlight this book about the fascinating world of ants. All readers will enjoy this book (part of a series by the author).

Lavies, Bianca. *Wasps at Home.* (New York: Dutton, 1991).
This book covers social wasps, paper wasps, and baldface hornets from a colony's beginnings in the spring to its demise in autumn. Full of great information.

Morris, D. *Insects that Live in Families.* (Milwaukee, WI: Raintree, 1987).
A thoroughly engrossing book detailing the behavior of bees, ants, and other insects that live in colonies.

Pringle, Lawrence. *The Golden Book of Insects and Spiders.* (Racine, WI: Western Publishing, 1990).
A beautifully illustrated book that includes diagrams and information about various insects and spiders.

Chapter 2

Great Crystal Bear

Carolyn Lesser

SAN DIEGO, CA: HARCOURT BRACE, 1996

Book Summary

Both powerful and gentle, polar bears thrive in the harsh conditions of the Arctic. Hunting for seals, quietly padding over ice floes, and curling up for a long winter's nap are but a few of the activities these wonderful giants engage in. This lyrical and poetically engaging book blends fact and fiction into a delightful and mystical tale about a single polar bear and his journey through a year. This is an incredible book that will open reader's eyes and ears to the magic of language and the enchantment of a tale well told.

Questions to Share

1. How was this description of a single animal similar to or different from other animal books you have read?
2. How did the illustrations help you learn more about this magnificent creature?
3. What was the most amazing fact you learned about polar bears?
4. How are polar bears different from other bears?
5. If you could, what would you like to share with the author of this book?

Major Book Project

This poetic book combines the best of fact and fiction to weave a story that captures readers' attention while providing them with incredibly fascinating information. Kids may be interested in combining fact and fiction through the use of the following activity.

In this activity, known as "Word Addition," children are provided with opportunities to blend a poetic phrase with a factual detail, two poetic phrases, or two factual details to create a unique combination. This process can be illustrated with the following example:

$$
\begin{array}{r}
\text{swirls of snow on the ice} \\
+ \text{ waving curtains of light} \\
\hline
\text{the vast Arctic night}
\end{array}
$$

In the example on the previous page, the phrase "swirls of snow on the ice" was added to the phrase "waving curtains of light" to create the addend "the vast Arctic night." Here are two more examples:

    swans take flight
+ daylight hours are shorter
_____

    fall is coming

    retreating sea ice
    snow geese huddled on nests
+ the sun rises higher
_____

    spring arrives

Invite youngsters to select some of their favorite phrases from the book and use them to create their own "Word Addition" problems. Youngsters may wish to begin by adding two phrases together and progress to the "addition" of three or four phrases to create unique and creative addends.

Provide opportunities for youngsters to share their "Word Addition" problems with others by collecting them in a notebook or by posting them on a large wall chart. Youngsters may also wish to trace and cut out an oversize outline of a polar bear from poster board and print their "Word Addition" problems across the face of this poster.

Special Project

The National Geographic Society (P.O. Box 96580, Washington, D.C. 20077-9964; 1-800-343-6610) produces a wonderful video entitled *Polar Bear Alert* (catalog no. 51290), which describes the annual northward trek of polar bears in the northern provinces of Canada.

If possible, obtain a copy of the video and show it to the youngsters with whom you work. Take time to discuss any similarities between the events portrayed in the film and those described in the book. You may wish to have youngsters complete a chart similar to the one below:

	Great Crystal Bear	**Polar Bear Alert**
Hunting/Eating		
Mating		
Traveling		

Discuss with children any differences between the events of the video and those described in the book. How accurate was the book in depicting the life of a solitary polar bear? What information was left out? What information should have been included? What would youngsters like to tell the author about polar bears that might be included in a future edition of the book?

Experiment

Obtain approximately two pounds of suet from your local butcher shop or meat dealer. Invite youngsters to make three different sized balls of suet. One ball should be 1 inch in diameter, another 2 inches in diameter, and the third 3 inches in diameter. Obtain three thermometers from a local drug store or variety store. Lie the thermometers side by side on a counter or table for an hour or so and check to be sure the temperature readings are relatively similar on all three.

Invite children to stick one thermometer into each of the three balls of suet so that the bulb is approximately in the center of each ball. Fill a basin, tub, or large pan with cold water and place lots of ice cubes in the water. Wait until the water is sufficiently cold (almost to the point of freezing).

Encourage children to record the temperature readings on each of the three thermometers. Each ball can then be placed in the container of ice water. Wait 15 to 20 minutes and invite youngsters to record the temperature readings on each thermometer again. What differences do they note between the original reading (room temperature) of each thermometer and the final reading (in ice water)? What differences between the three thermometers in the ice water are noted? Which one has the highest reading? Which one has the lowest reading?

Youngsters will note that the thermometer in the 3-inch ball of suet has a higher reading than the one in the 1-inch ball. The suet forms a protective barrier against the cold water. In polar bears, their blubber forms a protective barrier against the freezing temperatures of the Arctic. Their blubber, in combination with a thick coat of fur, provides them with a wonderful blanket of insulation.

Incredible Facts

🦌 **The polar bear is one of the largest meat-eating animals in the world.**
- *What are some other large meat-eating animals?*
- *What types of meat does a polar bear eat?*
- *What is the world's smallest meat-eating animal?*

🦌 **Polar bears are powerful swimmers and have been seen swimming 200 miles from land.**
- *Why would a polar bear swim so far out to sea?*
- *How far can you swim?*
- *Why do animals (including humans) swim?*

🦌 **A polar bear can run at speeds up to 19 miles per hour.**
- *What is the fastest you can run (in miles per hour)?*
- *What is the fastest animal in the world?*
- *What is the slowest animal in the world?*

🦌 **To avoid freezing, polar bears have a layer of blubber 4 inches thick.**
- *Where is the thickest part of your skin?*

- *What is the function of skin (in humans)?*
- *Why do we need to keep our skin clean?*

Related Literature

Brimmer, Larry. *Animals that Hibernate.* **(New York: Watts, 1991).**
A well-written introduction to a wide variety of hibernating animals throughout the world.

Chinery, Michael. *Questions and Answers about Polar Animals.* **(New York: Kingfisher, 1994).**
Lots of information in an easy-to-read format answers many kids' questions about the polar regions of the world.

Fair, Jeff. *Bears for Kids.* **(Minocqua, WI: NorthWord Press, 1991).**
A detailed and thorough overview of black bears in North America. Habits, diet, and behavior are presented.

Gill, Shelley. *Alaska's Three Bears.* **(New York: Paws IV, 1990).**
A blend of fact and fiction about the polar, grizzly, and black bears of Alaska.

Penny, Malcolm. *Bears.* **(New York: Bookwright Press, 1990).**
A simple, yet complete introduction to various bears found throughout the world.

Petty, Kate. *Bears.* **(New York: Gloucester Press, 1991).**
A mix of photos and color paintings acquaints readers with 12 different species of bears.

Sackett, Elisabeth. *Danger on the Arctic Ice.* **(Boston: Little, Brown, 1991).**
This book illustrates the dangers of life for harp seals in the Arctic.

Chapter 3

The Salamander Room

Anne Mazer

NEW YORK: KNOPF, 1991

Book Summary

A young boy discovers a little orange salamander in the woods and takes it home. Prodded by a series of questions from his mother, he thinks of all the imaginative ways he will care for his newfound friend and how his companion will live. His imagination, and the story, grows until its time for the boy and his friend to go to sleep.

Questions to Share

1. Have you ever had dreams like the boy in this story?
2. What are some things you must think about when caring for an animal?
3. How is the boy's room similar to or different from the natural living conditions of the salamander?
4. If you could choose any pet in the world, what would you have?
5. If you could add a sequel to this story, what would you write?

Major Book Project

Salamanders are wild creatures who belong in their natural environment. While this tale was a piece of fiction, it is important for youngsters to understand that wild animals should not be removed from their natural settings. This may confuse or disorient an animal or place it in an artificial environment in which it cannot search for adequate food, water, or shelter.

While children may be tempted to collect and examine salamanders at home or in the classroom, the following project provides them with ideas and tips on how they can locate and observe salamanders in the wild. Encourage children to maintain a journal or log of their observations and to follow-up with additional books or research materials in the school or public library.

Red Eft

This is the type of salamander illustrated in the book. The red eft salamander is typically found east of the Mississippi River and is distinguished by a red-orange body with a few small red spots on its back and sides. Its skin is rough and dry.

These salamanders grow up to $3\frac{1}{2}$ inches long with four short legs and a wormlike tail. They like to eat small worms and insects.

Youngsters can locate these creatures in the early fall, just prior to the first frost. They can be found under rocks and rotting logs. They may also be located in and around small ponds.

Tiger Salamander

The tiger salamander has a distinctive coloration—dark brown or black skin highlighted by white, gold, or greenish splotches on top and a yellow green belly. It has a wide head and a thick body.

Tiger salamanders grow to about 13 inches long with four short, stout legs. They enjoy eating worms, small insects, and mice.

These animals can be found in grassy or swampy areas. They like wet areas and often live near shallow ponds or near rocks and logs. One of the most distinctive features about these creatures is that the young occasionally eat each other.

Spotted Salamander

It's difficult to miss this creature with its black or dark brown body covered with yellow or orange spots. It also has a blue gray belly that is lightly speckled. It has four chubby legs.

Spotted salamanders reach a length of 6 to 10 inches long. Their diet consists of earthworms, small insects, snails, and spiders.

These animals can be located in and around small ponds, particularly those located in quiet areas of the woods or forest. They are particularly active at night. (*Note:* Children are cautioned not to handle this creature since its skin gives off a harmful liquid.)

Special Projects

A. Camouflage is the ability of an organism (plant or animal) to blend in with its surroundings. In doing so, the organism is able to escape detection and hide from its enemies. By the same token, it can conceal itself and sneak up on prey or lie unnoticed by other animals it wishes to turn into a meal.

The activity below helps children appreciate the value of camouflage to certain animals, such as salamanders.

Obtain 100 green toothpicks, 100 red toothpicks and a stopwatch or watch. Work with one or more children and mix up all the toothpicks together and spread them out on an area of lawn or grass approximately 25 feet by 25 feet. Give youngsters a time limit (one minute, two minutes, four minutes) to pick up as many toothpicks as possible. At the end of the designated time period invite children to note the total number of red toothpicks found in comparison with the total number of green toothpicks found. Invite youngsters to speculate on the reasons for different totals.

You may wish to explain to youngsters that they probably found more red toothpicks than green toothpicks because the green toothpicks were closer to the color of the test area than the red ones. Thus, animals that are able to blend into their surrounding through similarities in colors have a better chance of survival because of their natural camouflage. Animals who have distinctive colors may be at a disadvantage. Green salamanders, for example, are able to hide in a green environment more easily than brightly colored salamanders.

B. Provide youngsters with a set of index cards. On each card record a name of a plant or animal as in the following example:

> algae
> mosquito
> fish larva
> insects
> salamander
> osprey
> bacteria
> nutrients

The cards represent the line of succession in a food chain (algae —> nutrients). Provide opportunities for children to arrange the cards in a line indicating each organism and the other organisms that are dependent on it for their survival (youngsters may wish to attach lengths of yarn to the card to denote the relationships). Later, invite children to add additional cards to the deck and create additional relationships in the food chain. Additional library resources may be necessary.

Field Trip

Contact the biology or zoology department at a local college or university. Ask to be put in touch with a zoologist or herpetologist. Make arrangements for a visit to the campus so that children have an opportunity to talk with and ask questions of a "salamander expert." Prepare youngsters ahead of time by encouraging them to prepare a list of questions to ask the expert. After the trip invite the children to select one or more of the following projects based on the information they learned:

1. Invite a group of children to create a newsletter or newspaper describing the trip.
2. Encourage children to design a brochure on important points learned during the visit. The brochure can be distributed to other youngsters or attached as part of the book.
3. Youngsters may wish to put together a news broadcast about the trip and what was learned during the visit.

Across the Curriculum

Language Arts
Invite youngsters to research other books on salamanders (the list below is a good starting point). Encourage them to collect additional data about the lifestyles and habits of different species of salamanders.

After they have collected sufficient data, invite them to rewrite a portion of the story from the perspective of the salamander. What was observed? What was experienced? How did the salamander view the little boy? What did it think about living in the little boy's bedroom? How would its life be different if it lived in a house instead of living in the forest? Children may wish to include the answers to some or all of those questions in their brief rewrite of the story. Provide opportunities for children to share their stories with others.

❦ **Salamanders can regrow new tails, legs and feet.**
- *What other animals can regrow lost body parts?*
- *What is the advantage of regrowing lost body parts?*
- *Why can't humans regrow lost body parts?*

❦ **Most salamanders have four legs. Others have two legs, and a few have no legs at all.**
- *What would be the advantage of having no legs?*
- *What would be the advantage of having four legs?*
- *Do animals with more legs run faster than animals with fewer legs?*

❦ **A salamander known as the hellbender is able to breathe through lungs, gills, and also through its skin.**
- *What would be the advantages of having multiple ways of breathing?*
- *Would you want to have multiple ways of breathing?*

❦ **One species of blind salamander can either lay eggs or bear her young alive.**
- *What are some egg laying animals?*
- *What are some animals that give live births?*
- *Which group has the most number of species?*

Related Literature

Burns, Diane. *Snakes, Salamanders, and Lizards.* **(Minocqua, WI: NorthWord Press, 1995).**
A take-along guide that identifies and describes some of the most common reptiles and amphibians in the United States.

Parker, Nancy, and Joan Wright. *Frogs, Toads, Lizards, and Salamanders.* **(New York: Greenwillow, 1990).**
Blending fact and fiction, this book offers an engaging look at some of nature's most misunderstood creatures.

Wrigley, Robert. *Reptiles and Amphibians: Nature Stories for Children.* **(New York: Sterling, 1990).**
Lots of scientific information presented in a story format and highlighted by accompanying illustrations.

Chapter 4

Weird Walkers

Anthony D. Fredericks

MINOCQUA, WI: NORTHWORD PRESS, 1996

Book Summary

This book offers readers an amazing journey of discovery as they
learn about some of the most unusual animals on Earth. They meet a fish that
walks out of the water, a lizard that walks on the water, and a tree that "walks"
through the water. Information on protecting the environment of these special
creatures is also included.

Questions to Share

1. Which of the 12 animals in this book did you find to be most unusual?
2. Do you know of any other "weird walkers" that should be included in a book of this type?
3. What are some unusual means of transportation other animals use?
4. Which of the "Fascinating Facts" did you find most astounding?

Major Book Project

One of the most unusual creatures (also featured in this book) that travels the
Earth is the lowly snail. Among other amazing things, it is one of the few
animals that travels on its stomach—in fact, its stomach is also its foot!

The following activity provides youngsters with an opportunity to learn
more about snails in a controlled environment.

Find some land snails around the perimeter of your house (look in the
moist soil of gardens in the early morning hours). Invite children to put a 2-inch
layer of damp soil in a large clear jar and place the snails in it. Invite kids to place
some cheesecloth over the jar opening and fasten it with string or a large rubber
band (this will keep the occupants inside since snails can crawl up glass).

Instruct children to sprinkle the soil every so often to keep it wet and to
keep the jar in a cool shady place. Small pieces of lettuce can be placed in the
jar for food.

Encourage children to keep a journal or notebook of the snail's activities.
How fast does it move? How much does it eat in one day, one week, two
weeks? How active is it? How much does it travel?

Special Project

Invite children to measure the rates at which familiar animals travel. Using a stopwatch kids can time various insects, fish, mammals, or reptiles over a measured course and determine their speeds in miles per hour. A chart or graph of these relative speeds can be constructed for others to enjoy. Invite youngsters to record their own speeds, too. (How fast do they travel when walking? When running? When crawling?)

Write Away

Many of the animals mentioned in this book are endangered or live in environments that are rapidly dwindling. Unless action is taken soon, several of them (i.e., millipedes, sloths, tree frogs) will become extinct in the next few years.

Invite children to write to one or more of the following environmental agencies to obtain relevant literature on endangered species around the world. When the material arrives, invite children to make a listing of those animals that are most seriously imperiled, those that are endangered, and those that are threatened. Invite children to post their list (which can also be turned into an informative "Fact Sheet" or brochure to be shared with others) in the school or some other public building in town.

National Audubon Society
666 Pennsylvania Avenue SE
Washington, D.C. 20003

Friends of Wildlife Conservation
New York Zoological Society
185 Street, Southern Boulevard
Bronx Zoo
Bronx, NY 10460

Across the Curriculum

Math

Some of the creatures mentioned in this book travel quite rapidly (the ostrich, for example); others travel quite slowly (the sloth, for example). The rate at which an animal travels through its environment may be due to a number of factors, but chiefly revolve around its need to find food or protect itself from its enemies.

Invite children to make a large chart (on an oversized piece of poster board, for example) that lists the speeds at which selected animals travel. The chart can rank the order of animals from fastest to slowest or vice versa. Be sure to encourage kids to place animals on the chart with which they are very familiar (i.e., dogs, cats, guinea pigs, etc.). How much faster is their pet than the slowest animal on the chart? How much slower is the family dog than the fastest animal on the chart?

Experiment

Using a snail from the activity on page 15, invite children to predict the type of food a snail prefers. Place a piece of black construction paper on a table top. Have children place the snail in the middle of the paper and surround it with bits of food (i.e., pieces of lettuce, apple slices, pieces of celery, some cereal). Let go of the snail and watch what happens. Children may also wish to record how rapidly the snail travels toward its preferred food (How many minutes? What was the speed in miles per hour?).

🦌 **The mudskipper is a fish that walks across the beach and up the trunks of trees.**
- *What other animals live on the beach?*
- *How fast can a crab or other beach creature travel across the sand?*
- *Are there other animals that can breathe on land and in the water?*

🦌 **Frogs do not drink water—they absorb it through their skin.**
- *How much water should a person drink in one day?*
- *How far can a frog hop? How far can you hop? What's the difference?*
- *What's the most unusual frog in the world?*

🦌 **The linckia starfish is able to pull itself in separate directions until it breaks into two parts. Each of the two parts can grow into a new animal.**
- *Where are most of the world's starfish located?*
- *What is the most number of arms a starfish can have?*
- *What happens to a starfish when it breaks an arm?*

🦌 **Snails have up to 20,000 little teeth—all on their tongues.**
- *How many teeth does a normal human being have?*
- *Which animal has the largest number of teeth?*
- *Which animal has the fewest number of teeth?*

Related Literature

Dorros, Arthur. *Animal Tracks*. (New York: Scholastic, 1991).
Utilizing a guessing game format, the author introduces readers to animals and the tracks they leave behind.

Ling, Mary. *Amazing Crocodiles and Reptiles*. (New York: Knopf, 1991).
Clear accurate text highlighted with descriptive photography offers readers an inside look at some of natures most misunderstood creatures.

Myers, Jack. *Can Birds Get Lost? and Other Questions about Animals*. (Honesdale, PA: Bell Books, 1991).
Dozens of children's questions are accurately answered in a pleasant conversational tone.

Still, John. *Amazing Butterflies and Moths*. (New York: Knopf, 1991).
Lifelike photographs and sparse text make this book a delightful introduction to these remarkable insects.

Thompson, Ruth. *Creepy Crawlies*. (New York: Aladdin, 1990).
An interesting book loaded with inviting and detailed information about some strange critters.

Part II

Plants—
Ready, Set, Grow

Chapter 5

Miss Rumphius

Barbara Cooney

NEW YORK: VIKING/PENGUIN, 1982

Book Summary

Alice Rumphius desires to travel around the world and then to live by the sea—just like her grandfather. She also takes her grandfather's advice—to make the world more beautiful. This she does by scattering lupine seeds throughout the hills and valleys of her little town.

Questions to Share

1. What characteristics made Miss Rumphius an interesting character?
2. How is Miss Rumphius like one or more of your relatives?
3. What kinds of adventures away from home would you like to have when you get older?
4. If you were the author of this story, would you have written a different ending? Please explain.
5. What other things could Miss Rumphius have done to make the world more beautiful?

Major Book Project

Miss Rumphius made the world more beautiful by scattering lupine seeds throughout her community. Children can create some beautiful plantings, too—with their names!

Have children fill a large, flat cake pan with soil. Smooth it over so that the dirt is level. Moisten it with water and using the end of a toothpick or the tip of a knife, trace a child's name lightly into the soil.

Open a package of radish seeds and carefully plant the seeds in the grooves made for the letters of the child's name (make sure to follow the directions on the seed packet for proper planting depth and distance between seeds). Cover the seeds with soil and pat down lightly. Place the pan in a sunny location and water occasionally. After a few days the seeds will sprout into the shape of the child's name.

Invite youngsters to repeat this activity using different types of vegetable (i.e., parsley, mung beans) or flower (grass) seeds. Which varieties yield the "prettiest" names? Which varieties yield the most decorative or unusual names? Youngsters may wish to experiment with a wide variety of plantings.

Special Project

Invite youngsters to look in the local telephone book's yellow pages and create a list of all the services Miss Rumphius might have needed prior to her journeys. For example, a travel agent, a clothing store, a doctor (for immunizations), and a bookstore (for travel guides). Invite youngsters to suggest other possibilities. Encourage them to create a special set of yellow pages designed especially for Miss Rumphius. Information can be prepared with a word processing program and duplicated on yellow sheets of paper. The bound book can be shared with others.

Creative Dramatics

Invite students to create a prequel or sequel to the story. What events could have taken place before the story began? What events could have happened after the story concluded? Encourage youngsters to discuss changes they might have added to the story if the author had invited them to do so. Children may wish to speculate that they are a relative of Miss Rumphius. How would that fact affect the events that could happen before or after the book's story?

Across the Curriculum

Geography

Using a map of the world, invite students to write the names of the different places Miss Rumphius visited on small slips of paper. Post these slips around the outside of a wall map. Using lengths of yarn, encourage children to match the name of a specific location with its place on the map. Secure the two ends of each piece of yarn with a small piece of tape or pins.

Geography, Language Arts

Invite students to pretend that they are visiting (with Miss Rumphius) one or more of the countries mentioned in the book. Encourage them to create postcards (using 3 x 5 index cards) that they would send back to their families at home during the course of their travels. Youngsters may wish to visit a stamp store to obtain examples of postage stamps from the identified countries. Illustrations of the actual stamps can be drawn on the postcards.

Experiment

What do seeds need in order to begin growing? The following experiment can help youngsters discover the answer to that question. Invite children to label each of six sealable sandwich bags with a number (1–6). Cut three paper towels in half, moisten three pieces and place a piece in the bottom of each of three sandwich bags. Drop six radish seeds in each of the six bags and finish setting up all the bags as follows:

Bag #1: seeds, moist paper towel, no light (put in a closet or drawer), room temperature.

Bag #2: seeds, moist paper towel, light, room temperature.

Bag #3: seeds, dry paper towel, light, room temperature.

Bag #4: seeds, no paper towel, water (seeds floating), light, room temperature.

Bag #5: seeds, moist paper towel, no light, place in refrigerator or freezer.

Bag #6: seeds, moist paper towel, no light, room temperature, seeds covered by nail polish.

Invite children to record the date and time the activity was started and to check each bag twice daily for 2 to 3 weeks. Changes should be recorded in a journal or notebook. Invite children to make conclusions about the conditions necessary for seeds to germinate (begin growing).

Incredible Facts

✿ **There are more than 250,000 species of flowering plants in the world.**
- *How many different varieties of flowering plants grow around your house?*
- *What is the difference between a flowering plant and a nonflowering plant?*
- *What is the most unusual flowering plant in the world?*

✿ **The smallest seeds in the world belong to the begonia plant. It takes two million of them to weigh one ounce.**
- *What is the world's heaviest seed?*
- *How many seeds does a daisy, sunflower, or petunia produce in a season?*
- *What are some of the ways seeds travel?*

✿ **A winter rye plant can produce 380 miles of roots in 1.8 cubic feet of soil.**
- *Why are roots important to a plant?*
- *What plants have no roots whatsoever?*
- *You sometimes eat plant roots at home. Can you name some?*

✿ **Beans always climb a bean pole from right to left.**
- *How many different varieties of beans are there?*
- *What is the most unusual bean plant in the world?*
- *How many different ways to cook beans do you know?*

Related Literature

Ardley, Neil. *The Science Book of Things that Grow*. (New York: Harcourt Brace, 1991).
Using a photo-essay approach, this book explores many aspects of plant growth and includes step-by-step directions.

Markman, Erika. *Grow It!: An Indoor/Outdoor Gardening Guide for Kids.* **(New York: Random House, 1991).**
A comprehensive guide to indoor/outdoor gardening. Includes all aspects of cultivation and care as well as a host of investigative projects.

Miller, Suzanna. *Beans and Peas.* **(Minneapolis, MN: Carolrhoda, 1990).**
The history, description, and cultivation of beans and peas including colorful photographs.

Raftery, Kevin. *Kids Gardening: A Kid's Guide to Messing Around in the Dirt.* **(Palo Alto, CA: Klutz Press, 1989).**
Lively text and colorful illustrations highlight this informative guide for any young botanist.

Stidworthy, John Flowers. *Trees and other Plants.* **(New York: Random House, 1991).**
Using a question-and-answer format, this book offers lots of information about a wide variety of plants.

Chapter 6

Sky Tree

Thomas Locker

New York: HarperCollins, 1995

Book Summary

A tree stands on a hill by a river. The sky changes, the seasons change, and the tree goes through a series of marvelous transformations. A lyrical text complemented by wonderfully composed paintings highlight significant events in the year of a single tree. Questions at the bottom of each page provide readers with important events to think about and consider.

Questions to Share

1. Which of the illustrations made you feel the happiest, the saddest, the quietest, or the smallest?
2. What events in the life of this tree are similar to events in your own life?
3. Why do you think this tree is called the Sky Tree?
4. How is this tree similar to or different from other trees that you know?

Major Book Project

Invite children to select a tree near where they live. If possible, a deciduous tree (one that loses its leaves each year) should be chosen. The tree should be in a place that is easy to reach, because it will be visited for several weeks or several months.

To start, invite youngsters to draw a picture (or take photographs) of their "adopted" tree. Encourage them to note any unusual markings, characteristics, or patterns. These observations can be written in an ongoing journal and supplemented with photographs and illustrations.

Invite children to measure three feet up the tree trunk from the ground and tie a piece of string around the trunk at that point. Have children measure the section of string to determine the circumference of the tree. This measurement can be taken periodically over the next several weeks or months to denote any changes.

Invite children to collect some of the tree's leaves and save them in sealable plastic sandwich bags. This, too, can be done periodically throughout the year. Have kids place a piece of tracing paper or copy paper against the bark of the tree and rub a crayon over the paper until the pattern of the tree appears.

Encourage youngsters to keep a "running record" of the various animals that visit the tree. These may include birds, insects, small mammals, reptiles and other live organisms. Which type of animal seems to "visit" the tree most often? Is the tree a home for any type of animal?

Youngsters should be encouraged to persist with this project over an extended period of time—a year would be ideal. Take time periodically to discuss the changes the tree experiences during the course of the year—How does it grow? How does it react to the seasons of the year? What kinds of visitors does it get? Children should maintain an ongoing journal on the life of their "adoptee."

Special Projects

A. Trees provide humans with many different products that we use every day. Often, we may take for granted the role that trees play in our everyday lives. Kids may feel the same way about trees as well.

Invite children to initiate a chart similar to the one below. Using library resources, encourage them to add to the items already included on the list. Take time to discuss with them the various uses trees have in your own life, your family life, or around the house. This list can be added to over a long period of time.

How We Use Trees			
Food	*Medicine*	*Industrial Products*	*Home Products*
cocoa	quinine	rubber/dyes	lumber/paper
cinnamon			
maple syrup			

B. Talk with children about the idea of forming a "Tree Patrol Club." This club would be made up of members who are interested in doing something to protect and preserve trees in the local neighborhood. Members would actually "patrol" their neighborhoods, looking for things that might create problems for trees such as garbage, toxins, pollution, and so on. Once members have identified one or more problems for one or more trees, they could take a photograph of the problem, illustrate it, or write about it. Periodically, members can share what problems they observed and, as a group, problem solve what needs to be done to eliminate a particular problem. Then they can take whatever steps are necessary to reverse the problem. This may require the assistance of other people or perhaps writing to local, city, county, or state officials or agencies. Group members should keep a log/scrapbook of what the club does to help preserve trees.

Write Away

There are a number of organizations and groups that provide brochures and other information on trees to the public for free (or for a very small charge). Invite youngsters to write to one or more of the following groups and ask for pertinent literature:

Forest Service
U.S. Department of Agriculture
P.O. Box 96090
Washington, D.C. 20090
(Ask for the poster, "How a Tree Grows")

U.S. Geological Survey
P.O. Box 25286
Denver, CO 80225
(Ask for the brochure, "Tree Rings: Timekeepers of the Past")

The American Forestry Association
Global ReLeaf Program
P.O. Box 2000
Washington, D.C. 20013

National Arbor Day Foundation
100 Arbor Avenue
Nebraska City, NE 68410

What's Cooking?

Several of the illustrations in this book show the tree in the spring and summer—times when people are outdoors enjoying nature. Certainly one of the activities people often engage in is having picnics beneath the branches of a large tree. Invite children to make one or more of the recipes below and to plan a picnic under the branches of a nearby tree. (*Note:* Other recipes may be added to the ones suggested below.)

Corn Muffins

Ingredients

1 cup yellow cornmeal
1 cup flour
2 tablespoons sugar
4 teaspoons baking powder
1/2 teaspoon salt
1 cup milk
1/4 cup shortening
1 egg

Directions

Heat the oven to 425°F. Grease the bottoms of 25 medium muffin cups. Blend the ingredients for about 20 seconds in a large mixing bowl. Fill the muffin cups about $^2/_3$ full. Bake for 15 minutes, immediately remove from the pan.

Carrot Sunshine Salad

Ingredients

1 cup boiling water
1 package lemon flavored gelatin
1/2 cup cold water
1/8 teaspoon salt
1 can crushed pineapple
1/2 cup shredded carrot

Directions

Empty the gelatin into a small bowl and mix with the boiling water. Stir until the gelatin is dissolved. Stir in the cold water, salt, and pineapple (with syrup). Chill until the mixture is slightly thickened. Fold the carrot into the mixture. Pour the mixture into a mold and chill until set. Cut into squares and serve.

Arts and Crafts

Invite children to obtain a piece of cardboard (approximately 8½ x 11 inches). Cut a rectangle out of the center, leaving a border of approximately 1½ inches. Encourage them to hold the cardboard "frame" at arm's length to "frame" a nearby tree. What do they see? Where is the light? What does the sky look like behind the tree? Is the top of the tree taller than the trunk? Is the tree wider than it is tall? Does it look like a triangle, a square, a rectangle, or a circle?

Based upon their observations, encourage them to talk about how they might illustrate the tree. Provide them with simple materials (crayons, paper, etc.) and invite them to draw a rendition of their tree. How do they see a tree differently than the illustrator of this book? Take time to discuss any differences or similarities in illustrations that exist among a group of children (i.e., choice of tree, drawing style, color, etc.).

Incredible Facts

❧ **A giant sequoia's roots can cover more than an acre of land, yet are only 3 to 5 feet below ground surface.**
  • *What tree has the longest roots?*
  • *What tree has roots that spread out the farthest?*
  • *Why do trees have roots?*

❧ **Tree rings also record earthquakes.**
  • *How do scientists use tree rings to determine the age of a tree?*
  • *Why are some tree rings wide while others are thin?*
  • *Do all trees have tree rings?*

✤ **The typical tree receives about 10 percent of its nutrition from the soil. The rest comes from the atmosphere.**
  - *What are some nutrients trees need to survive?*
  - *How do pollutants affect the growth of a tree?*

✤ **A typical apple tree loses up to 20 quarts of water a day.**
  - *How do trees "breathe"?*
  - *How many gallons of water a year does a typical apple tree lose?*
  - *Which state grows the most apples?*

Related Literature

Arnosky, Jim. *Crinkleroot's Guide to Knowing the Trees*. (New York: Bradbury, 1992).
An easy-to-read handbook with clear illustrations of leaves, seeds, needles, and seed cones.

Collard, Sneed. *Green Giants*. (Minocqua, WI: NorthWord Press, 1994).
An examination of twelve of Earth's tallest trees from coastal redwoods to the Tualang tree of Malaysia.

Reed-Jones, Carol. *The Tree in the Ancient Forest*. (Nevada City, CA: Dawn Publications, 1995).
A repetitive, cumulative verse that depicts the remarkable web of plants and animals that live on a single fir tree.

Thornhill, Jan. *A Tree in a Forest*. (New York: Simon and Schuster, 1992).
The 212-year life cycle of a sugar maple tree. The emphasis is on the flora and fauna that depend on the tree for their existence.

Viera, Linda. *The Ever-Living Tree*. (New York: Walker, 1994).
The life story of a redwood tree over a period of nearly two thousand years.

Chapter 7

The Tree in the Ancient Forest

Carol Reed-Jones
NEVADA CITY, CA:
DAWN PUBLICATIONS, 1995

Book Summary

The remarkable web of plants and animals living around a single old fir tree takes on a life of its own in this stunningly illustrated story. The repetitive, cumulative verse aptly portrays the amazing ways in which the inhabitants depend on one another for survival. A guide to forest creatures and their interrelationships is also included as a learning tool. A magnificent book of discovery.

Questions to Share

1. What new forest animals did you learn about in this book?
2. Why do plants and animals need each other for survival?
3. What is the most important plant in or near your house?
4. What are some of the various ways that humans use trees?
5. What are some dangers to the trees of the world?

Major Book Project

Invite youngsters to create their own cumulative poetry in a style similar to that in this book. Encourage them to select a tree, flower, bush, or some other plant in or around their home and develop an appropriate story-poem. Following are two examples of cumulative poems you may wish to share with children:

This is the Smith house.

This is the great green peach tree
That grows near the Smith house.

These are the fruits that hang from the branches
That encircle the great green peach tree
That grows near the Smith house.

These are the leaves all shiny and long
That brush against the fruits that hang from the branches
That encircle the great green peach tree
That grows near the Smith house.

⚘ ⚘ ⚘

This is Marla's bedroom.

This is the potted plant
That rests inside Marla's room.

This is the brown speckled bowl
That holds the potted plant
That rests inside Marla's room.

This is the tall sunny window
That brightens the brown speckled bowl
That holds the potted plant
That rests inside Marla's room.

Provide opportunities for children to compose several different examples of these cumulative story-poems for different examples of various plants. These poems can be collected into a journal or notebook and shared with others.

Special Project

A. Invite youngsters to plant a tree in their yard, a local park or playground area, or alongside a nearby road. Visit a local nursery or gardening center. Talk with the people who work there to figure out what kind of tree to plant. Talk with children about the factors that will help them select the most appropriate tree(s). These may include initial cost, care and maintenance, location, climate, availability of nutrients, soil conditions, long term needs, aesthetic value, and whether the tree is native to the local area.

Follow appropriate planting instructions for the tree (the people who work at the gardening center or nursery can help you out). Invite children to initiate a journal on the life of the tree in its new location. They can record its height, diameter, width, condition, number of leaves, number of branches, and other pertinent data. Encourage them to keep a running record of the tree over an extended period of time (i.e., several months, a year, several years).

Later, youngsters may wish to talk with friends, neighbors, or other people in the community about starting a community tree-planting effort. Encouraging each family in a neighborhood to plant just one new tree can be an important community effort.

B. Invite children to gather information and data from the school and/or public library. They can put together a booklet or notebook entitled "Tree Olympics"—a compendium of the "world records" held by individual trees, varieties of trees, or groups of trees throughout the world. The following "world records" may help get them started:

World's tallest tree:
World's oldest tree:
World's smallest tree:
World's widest tree:
Tree with the longest roots:
Tree with the biggest seeds:
Tree with the smallest seeds:
Tree with the widest leaves:
Most common tree:
Rarest tree:

The information can be researched and gathered together in a journal or notebook for sharing with others.

Write Away

The following organizations and groups offer loads of information and important data on how to plant and grow trees. Invite youngsters to request relevant brochures and documents.

TreePeople
12601 Mulholland Drive
Beverly Hills, CA 90210

Worldwatch Institute
1776 Massachusetts Avenue NW
Washington, D.C. 20036

American Forestry Association
Global ReLeaf Program
P.O. Box 2000
Washington, D.C. 20013

National Arbor Day Foundation
Arbor Lodge 100
Nebraska City, NE 68410

When material from these groups arrives, invite youngsters to organize it into various categories. These could include one or more of the following:

Growing Tips, Forests and Forestry, Saving Our Forests, Growing a Tree, North American Trees, How to Care for a New Tree, and Trees and Our Future.

❀ **The average American uses seven trees a year in paper, wood, and other products made from trees.**
- *How many wood products are in your home?*
- *How much paper does your family use each week?*
- *How would humans get along if there were no trees?*

❀ **The tallest tree in the world is a California redwood, which measures 360 feet tall.**
- *How tall are you?*
- *How much taller than you is the tallest tree?*
- *How many of you would you need to stack one on top of the other to reach the top of the tallest tree?*

❀ **The fastest growing tree is the Albizzia falcata tree of Malaysia, which grows as much as 33 feet a year.**
- *How much did you grow last year?*
- *How much have you grown since you were born?*
- *Do plants or humans grow the fastest?*

❀ **There is a bristlecone pine tree growing in eastern California that is more than 4,600 years old.**
- *What is the oldest tree in your yard or neighborhood?*
- *How do scientists measure the age of a tree?*
- *Why would it be important to know the age of a tree?*

Related Literature

Burnie, David. *Tree*. (New York: Knopf, 1988).
Lots of facts and clear descriptive photographs highlight this book on the life cycles of trees.

Hiscock, Bruce. *The Big Tree*. (New York: Atheneum, 1991).
The social history encompassed within a sugar maple's life span from Revolutionary War days to the present.

Markle, Sandra. *Outside and Inside Trees*. (New York: Bradbury, 1993).
Clear writing explains all the parts of a tree and how those parts are used.

Quinn, Greg. *A Gift of a Tree*. (New York: Scholastic, 1994).
There are seeds and instructions for growing a tree along with information on the importance of trees to humans.

Thornhill, Jan. *A Tree in a Forest*. (New York: Simon and Schuster, 1992).
The 212-year cycle of a sugar maple tree and its interrelationships with the flora and fauna around it.

Part III

Environment—
Clean and Green

Chapter 8

The Great Kapok Tree

Lynne Cherry

NEW YORK: GULLIVER BOOKS, 1990

Book Summary

A young man enters the rainforest to cut down a Kapok tree, but before he knows it the heat makes him tired and weak. The man sits down to rest and falls asleep. While he sleeps the animals of the forest whisper in his ear not to cut down the Kapok tree. Each animal has a different reason for preserving the tree. Upon awakening, the man realizes the importance of the Kapok tree.

Questions to Share

1. If you could be one of the animals in the story, which one would you be? Why?
2. What do you think would happen if all the rainforests were destroyed?
3. What can kids do to protect the rainforests?
4. Why do some people do things that are harmful to the environment, such as littering?
5. Why do you think the author decided to write a book about the Amazon rainforest?

Major Book Project

There are several environmental organizations in the United States working to preserve the rainforests of the world. Invite children to write to these groups and ask for information, newsletters, brochures, and facts about the world's rainforests and the efforts to protect these valuable areas. Here are some groups you may wish to contact:

Children's Rainforest
P.O. Box 936
Lewiston, ME 04240

Rainforest Action Network
450 Sansome Street, Suite 700
San Francisco, CA 94111

Rainforest Alliance
65 Bleeker Street
New York, NY 10012-2420

Rainforest Preservation Foundation
P.O. Box 820308
Ft. Worth, TX 76182

Save the Rainforest
604 Jamie Street
Dodgeville, WI 53533

After information arrives, invite children to review it and assemble a descriptive display (for the home, the classroom, or a local business) that presents some of the most important facts and features on the rainforest along with methods and ways to conserve this ecosystem. Invite them to compare this data with that presented in children's literature (see below).

Field Trip

Look in the local phone book for the nearest recycling center (paper, aluminum, or glass). Make arrangements to take the children to the center and observe the operation. It may be helpful to ask children to generate a list of possible questions before the visit. Children should be aware that recycling helps preserve the environment as well as preserves plant and animal life (children may be interested in learning that sixty thousand trees are needed for just one run of *The Sunday New York Times*).

After the visit invite children to discuss the need for additional recycling efforts in their community or city. Is a letter writing campaign necessary to have city officials mandate recycling? How can local citizens become involved in recycling efforts? What can kids do to promote recycling in their local area?

Write Away

Children may wish to adopt an animal—especially an animal from the rainforest. Invite kids to write to the American Association of Zoological Parks and Aquariums (4550 Montgomery Avenue, Suite 940N, Bethesda, MD 20814) for information on "adopting" an animal, particularly an endangered species. When the information arrives, allow kids to decide on what type of animal to adopt and to create posters or illustrations of the "adoptee" to display for others.

Across the Curriculum

Math

Kapok trees are some of the largest trees in the rainforest. Often reaching heights of 100 feet or more, they frequently tower over the tops of more numerous trees that inhabit the forest.

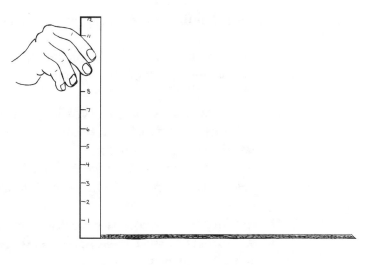

Invite children to measure the heights of various trees in their neighborhood. Each child should have two rulers and a length of string. Go outside on a sunny day and locate a nearby tree. Have each child stand a 12-inch ruler on the ground (see illustration) and measure the length of the ruler's shadow. Have each child take a length of string and measure the length of the shadow of a nearby tree. Invite children to use the following formula to compute the exact height of the tree:

$$\frac{\text{height of ruler}}{\text{length of ruler's shadow}} \quad X \quad \frac{\text{height of tree}}{\text{length of tree's shadow}}$$

For example:

Height of ruler = 12 inches
Length of ruler's shadow = 24 inches
Length of tree's shadow = 720 inches

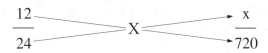

12 X 720 = 8,640
24x = 8,640
8,640 ÷ 24 = x
x = 360 inches (30 feet)
The tree is 30 feet tall.

♻ **Every second of every day a piece of rainforest the size of a football field disappears.**
  • *How large is a football field?*
  • *How many football fields of rainforest are lost in an hour?*
  • *How many football fields of rainforest are lost in a day?*

♻ **Some tropical rainforests may get as much as 400 inches of rain every year.**
- *How many feet of rain do some rainforests get a year?*
- *How many yards of rain do some rainforests get a year?*

♻ **The anaconda snake, the world's largest, lives in the rainforests of South America. It can grow to more than 30 feet in length and weigh over 330 pounds.**
- *How many inches long can an anaconda grow?*
- *How many ounces does the largest anaconda weigh?*

♻ **The largest flower in the world—the rafflesia—lives in the rainforests of southeast Asia. It is three feet across and weighs up to 36 pounds.**
- *How much larger is the rafflesia flower in comparison to a daisy?*
- *How much bigger is the rafflesia flower in comparison to a sunflower?*

Related Literature

Baker, Jeannie. *Where the Forest Meets the Sea.* **(New York: Greenwillow, 1988).**
Kids will delight in the magical and inventive illustrations used to tell the story of a disappearing Australian forest.

Cowcher, Helen. *Rainforest.* **(New York: Farrar, 1988).**
This books raises ecological issues through a well-told tale and wonderful illustrations.

Fredericks, Anthony D. *Exploring the Rainforest.* **(Golden, CO: Fulcrum Publishing, 1996).**
Dozens of engaging activities highlight this hands-on examination of a dwindling ecosystem.

George, Jean C. *One Day in the Tropical Rainforest.* **(New York: HarperCollins, 1990).**
The struggle over land in the Amazon rainforest is told through the eyes of a young Indian boy. A "must read" for young environmentalists.

Landau, Elaine. *Tropical Rainforests Around the World.* **(New York: Watts, 1990).**
A thorough introduction to rainforest life and the implications when this valuable ecosystem is destroyed.

Silver, Donald. *Why Save the Rainforest?* **(New York: Messner, 1993).**
A wonderful survey of life in the rainforest and the balance of nature that is in danger of being eliminated. Includes rich and colorful illustrations.

Chapter 9

Mother Earth, Father Sky

Selected by Jane Yolen

HONESDALE, PA:

BOYDS MILLS PRESS, 1996

Book Summary

This book is a delightful anthology of poems about the Earth we live on. It celebrates the diversity of life, the wonders of nature, and the magnificent interrelationships that exist between plants and animals. So, too, does it decry the ways in which humans harm their own planet—how pollution, deforestation, and ecological degradation are slowly sickening the environment—ours as well as that of the organisms with whom we share this planet. A perfect addition to any study of the environment, this book offers youngsters unique perspectives and lyrical insights into the global habitat we all share.

Questions to Share

1. Which of the poems did you enjoy the most? What aspects of that poem did you like?
2. Which of the poems made you the angriest? Which one made you mad at some of the things humans are doing to the environment?
3. Why is poetry sometimes more powerful in making a point—such as our need to preserve the planet?
4. Which of these poems would you want to send to all the leaders of the major countries of the world? Please explain.
5. What did you find distinctive or unusual about the illustrations in this book?

Major Book Project

Following are four environmental activities. Invite children to select one or more of these projects and complete them. Afterward, ask youngsters to identify one or more poems in the book that match each activity. Why does a particular poem or group of poems "connect" with a specific project? What elements of a particular poem are reflected in the "learnings" of a specific activity? Plan opportunities for youngsters to discuss the "matchings" they make and the rationale they used in making those comparisons. Of course, it will be important to emphasize that there are no right or wrong answers and that children should feel free to interpret a poem (or an activity) according to their own experiences or viewpoints.

A. Invite youngsters to add ¼ teaspoon of dry blue tempera paint to ½ cup of dry soil and mix thoroughly. Set a kitchen funnel in the mouth of a wide mouth jar and place a coffee filter in the funnel. Pour the soil/paint mixture into the filter. Have children pour ½ cup of water into the funnel and notice the color of the water running into the jar. Pour off the colored water and replace the funnel. Have children repeat the activity by pouring another ½ cup of water into the funnel. Repeat several more times. (The blue tempera paint represents soil nutrients. However, when there is a lot of rain or lots of soil erosion, those nutrients are washed away leaving a nutrient-poor soil. Plants find it difficult to grow in this soil.)

B. Fill four plastic drinking cups with potting soil. Invite children to plant 5 to 6 bean seeds in each of the cups, and water according to the package directions. Mix up some plaster of paris (available at most variety or arts-and-crafts stores) and pour a layer 2 inches thick over the top of two of the plastic cups. Place all four cups on a window sill or in a sunny location. Keep the soil moist in the two "non-plastered" cups. Invite children to note what happens in each of the cups over a period of 2 to 4 weeks. (*Note:* The seeds in all four cups will begin to germinate. However, the new plants in the "plastered" cups will not be able to continue their growth because plaster prevents them from sprouting above the surface. The other two plants will be able to proceed with their normal life cycle. When seeds or plants are covered by concrete or asphalt it is difficult for them to grow normally.)

C. Obtain several paint paddles (wooden sticks used to stir cans of house paint—available at all hardware stores). Invite youngsters to stick some two-sided cellophane tape on one side of several different paddles. Invite them to stick the paddles into the ground at various locations around their homes. Paddles should face several different directions (North, South, West, East). At regular intervals (once or twice a week) invite children to observe the paddles and record the amount of dust, dirt, or soil sticking to the tape. (Children will note that selected paddles collect more dust and dirt than others. As the wind blows it picks up this dust and scatters it. Depending on the wind's direction [N, S, W, E] the paddle[s] facing that direction will have collected the most dust. The amount of dust on a particular paddle is also a rough measure of the amount of wind erosion in that area.)

D. Obtain several different varieties of citrus fruit (oranges, lemons, grapefruit). Cut them in half. Invite youngsters to scoop out the insides and make four small holes around the perimeter of each half. Tie pieces of string to the holes around each fruit. Fill each one with a bird seed mixture, and hang them in nearby trees or bushes. Invite children to note the different varieties of birds that visit these feeders. (These "Avian Snack Bars" provide children with wonderful opportunities to observe wildlife [birds] in their natural environment. Careful observation will reveal the types of birds native to your area of the country. By experimenting with different mixtures of bird food, children will learn about the dietary preferences of specific species.)

Special Project

The three "R's" of environmental conservation are Reduce, Reuse, and Recycle. Invite youngsters to create notebooks or prepare large sheets of paper with these three works as the headings or titles. Encourage them to select poems from the book that support the concept of Reduce, Reuse, and Recycle and to write the titles of the representative poems in the appropriate section. As an alternative activity, encourage children to write the titles of poems that are in opposition to each of the three R's.

Provide opportunities for children to discuss the selection and placement of poems in each category. What elements of a particular poem caused it to be categorized as a "Reduce," "Reuse," or "Recycle" poem? Were there any poems that could not be placed in any of these three groupings? Is there a collection of poems that could be placed in an entirely different category? If so, what would be the name of that grouping?

Field Trip

Share with children one or more of the following poems from the book: "At the Dark's Edge," "The Bulldozer," "To Look at Anything," and "The Samara." Discuss with them some of the images these poems evoke. How do the poems make them feel? Do the poems move them to action? Do the poems make them happy? Sad? Angry?

After discussing one or more of the poems take the children you work with outdoors for a walking field trip of the neighborhood or local community. Invite children to look for the sights and sounds presented in any one of those poems. Are there parts of the neighborhood in which trees are being planted ("The Samara")? Are there places in the community where leaves are blowing about ("To Look at Anything")? Are there construction sites in which plants have been pulled up or animals displaced ("The Bulldozer")?

Plan a follow-up session with children on the feelings they have while reading or listening to a particular poem and how those feelings may be illustrated in the neighborhood or community. Did any poem make you more sensitive to what is taking place in your own backyard or neighborhood?

As an additional follow-up activity invite youngsters to create their own poetry for situations or conditions in the neighborhood for which there is not a poem or group of poems in the book. For example, there is not a poem about ocean pollution. If children live near the shore and have seen garbage or other oceanic pollution wash up on shore how does that make them feel? Can they develop their own poem to express their feelings about that specific situation? Poems in the book can be used as appropriate models.

Music and Movement

Several of the poems in this book lend themselves to dance interpretations, rhythmic movements, and artistic impressions. Invite youngsters to each select an individual poem. Encourage them to create an artistic presentation of the poem through some type of dance or musical exhibition. These can be very brief interpretations by single children or slightly more complex with a small group of individuals working together. For example, here is a possible interpretation of the poem "Oh World, I Wish." (Just the first few lines are presented.)

Oh World,	*(sweep both hands upward)*
I wish you were my mother,	*(bring hands inward and wrap around body)*
For I would spread my fingers out	*(spread fingers out)*
Against your earth face	*(bring hands slowly inward to face)*
And smell again the good brown smell.	*(hold hands close to nose)*

♻ **The average American family produces about 100 pounds of trash every week.**
  • *How much trash does your family produce in a week?*
  • *How much trash does your family produce in a year?*
  • *What happens to all the trash your family throws away?*

♻ **By recycling one ton of paper, 17 trees are saved.**
  • *What are some recycling efforts your family practices?*
  • *Why don't more people recycle?*
  • *What are some other advantages of recycling?*

♻ **Each year, about 2.3 trillion gallons of liquid wastes are discharged directly into U.S. coastal waters.**
  • *What are some of the long-term effects of that discharge?*
  • *How does all that waste affect you directly?*
  • *What are some of the effects on ocean life?*

♻ **In the United States alone, more than 500,000 trees are chopped down to supply paper for one week's worth of newspapers.**
  • *How many trees are chopped down in one year?*
  • *Where do all those trees come from?*
  • *How can we be sure there will be enough trees for the future?*

Related Literature

Arnosky, Jim. *In the Forest: A Portfolio of Paintings*. (New York: Lothrop, 1989).
The beauty and majesty of an important ecosystem are wonderfully illustrated in this book.

Baker, Jeannie. *Where the Forest Meets the Sea*. (New York: Greenwillow, 1988).
Readers will delight in the magical and inventive illustrations used to tell the story of a disappearing Australian forest.

Ballamy, David. *The Roadside*. (New York: Clarkson Potter, 1988).
This book examines one particular aspect of our environment and the responsibilities of humans to preserve it.

Dekkers, Midas. *The Nature Book: Discovering, Exploring, Observing, Experimenting with Plants and Animals at Home and Outdoors*. (New York: Macmillan, 1988).
The title says it all. This book provides youngsters with active opportunities to utilize all the processes of science in learning about nature.

Huff, Barbara. *Greening the City Streets: The Story of Community Gardens*. (New York: Clarion, 1990).
This book presents a wonderful look at some citizen's efforts at revitalizing an urban environment.

Rylant, Cynthia. *Night in the Country*. (New York: Bradbury, 1986).
A rich and wonderfully created examination of country life at night.

Van Allsburg, Chris. *Just a Dream*. (Boston: Houghton Mifflin, 1990).
A young boy discovers the need to preserve trees in this wonderfully illustrated tale of recycling and renewal.

Chapter 10

A River Ran Wild

Lynne Cherry

San Diego, CA: Gulliver, 1992

Book Summary

A beautifully illustrated book, this story concerns the "life history" of the Nashua River in New England. The people who lived along its banks and how their customs, traditions, and culture were shaped by the forces and resources of the river are wonderfully told. Then, the people overpowered the river—filling it with pollution and toxins. However, the actions of several individuals helped return the river to its original state. This book is a "must-have" for any family or classroom—a marvelously told tale of environmental concern and action.

Questions to Share

1. Is the Nashua River similar to any river, stream, or water source near where you live?
2. Do you believe that people intentionally set out to pollute a river? Explain.
3. How did the border illustrations (on the left-hand pages) help you learn more about the people who lived near the river?
4. As you look at the various illustrations of the river throughout the book, the artist uses many different colors. Why do you think she did that?
5. Do you know of any other examples in which one person made a difference?

Major Book Project

Invite children to visit a nearby stream or river. Encourage them to look carefully at the water as it flows by. Invite them to dip an empty plastic container in the water and raise it to their noses to smell. Ask them to sniff the air. Ask them if they noticed any of the following:

- A rotten egg smell. Streams with this odor are heavily polluted with sewage being dumped into the water.

- A shiny or multicolored film on the water surface. Streams that look like this usually indicate that oil or gasoline have been dumped into them.
- The water color is very green. These waterways probably have a lot of algae growing in them. Too much algae means there isn't enough oxygen in the water for fish and plant life.
- Foams or suds floating in the water. This usually means that detergents or other soapy wastes are leaking into the water from nearby factories or homes.
- Cloudy or extremely muddy water. This means that there are large amounts of mud, dirt, or silt in the water. It also means that plants and animals may not be getting enough oxygen. Another cause may be soil erosion taking place upstream.
- Bright colors, such as orange or red, on the surface. This is usually a sign that pollutants are being released into the stream by factories or industries upstream. Inform children that clean, clear, and transparent water that has no smell to it is the best kind of water for aquatic plants and animals and for humans, too.

If a polluted water source is detected in your area, invite youngsters to select one or more of the following actions to raise public awareness about the polluted waterway.

1. Invite them to write a letter to the editor of the local newspaper.
2. Create a series of posters or signs for posting in local businesses.
3. Prepare a brief public service announcement (a local advertising agency can assist you with this) for distribution to local radio stations.
4. Write a letter of complaint to a local or state legislator informing that individual of the polluted waterway.
5. Ask the local newspaper to insert a flyer or leaflet in an edition of the daily paper.
6. Create an inexpensive leaflet and ask that it be placed in the grocery bags of patrons at the local supermarket.
7. Have them send a letter of concern to the local chamber of commerce.

Encourage youngsters to brainstorm for additional activities and outreach efforts. What other ways can they think of to inform the public and inspire the public to do something about a polluted or endangered waterway?

Write Away

Invite youngsters to write to the following organizations to request information and data on river and stream protection:

The Isaak Walton League of America
1401 Wilson Boulevard, Level B
Arlington, VA 22209

Invite kids to request a copy of their free *Save Our Streams* booklet as well as other information.

Renew America
1400 16th Street, Suite 710
Washington, D.C. 20036
They have a collection of environmental success stories about kids who have made a difference.

20-20 Vision
69 South Pleasant Street, #203
Amherst, MA 01002
This group can locate the address of any government official in the country—great if kids would like to target their letters to specific bureaucrats.

Keep America Beautiful
99 Park Avenue
New York, NY 10016
This organization prepares a brochure entitled "Pollution Pointers for Elementary Students," which is a list of environmental improvement activities.

Across the Curriculum

Social Studies, Language Arts

Invite children to look at the border illustrations on the left-hand pages. Each drawing is related to the time period described in the accompanying text about the river.

Provide opportunities for students to investigate (through various library resources) one or more of the border drawings and their relationship to the time frame of the story on that page. For example, on one page the author talks about the "start of the new century" and the various things that were taking place on the Nashua River during those times. The border illustrations on that page include the first telephone, an Edison phonograph, the first bicycle, the first camera, and the first sewing machine, for example. Invite youngsters to research why these items were important to that time period. Why did the artist/author select these particular items? What other items could be included on this page? What items are more representative of that time period?

After sufficient discussion, encourage youngsters to compose a brief story (two paragraphs) about the time in which they live. Invite them to create a series of border illustrations (like in the book) that would depict items, features, or events associated with their specific time period. What items would they choose? Which ones should not be included? If you are working with a group of children, you may wish to have them combine their stories and illustrations into a book similar in design to *A River Ran Wild*.

Experiment

The following experiment will help youngsters learn about various forms of pollution and its effects on the environment. This activity *requires* adult supervision and guidance.

Provide children with four empty and clean baby food jars. Label the jars "A," "B," "C," and "D." Fill each jar halfway with aged tap water (regular tap water that has been left to stand in an open container for 72 hours). Put a $1/2$-inch layer of pond soil into each jar, add one teaspoon of plant fertilizer, then fill each jar the rest of the way with pond water and algae. Allow the jars to sit in a sunny location or on a window sill for two weeks.

Next, work with youngsters to treat each separate jar as follows:

Jar "A": Add 2 tablespoons of liquid detergent.
Jar "B": Add enough used motor oil to cover the surface.
Jar "C": Add $1/2$ cup of vinegar.
Jar "D": Do not add anything.

Allow the jars to sit for four more weeks.

Children will notice that the addition of the motor oil, vinegar, and detergent prevents the healthy growth of organisms that took place during the first two weeks of the activity. In fact, those jars now show little or no growth taking place, while the organisms in jar "D" continue to grow.

Detergent, motor oil, and vinegar are pollutants that prevent organisms from obtaining the nutrients they need to grow.

- The detergent illustrates what happens when large quantities of soap are released into a waterway.
- The motor oil demonstrates what happens to organisms after an oil spill.
- The vinegar shows what can happen to organisms when high levels of acids (i.e., acid rain) are added to a pond or stream.

Children will be able to see the effects of various pollutants in their miniature ecosystems. Plan time to discuss the implications of large quantities of these pollutants discharged into a waterway over a long period of time. What would be the short- and long-term consequences of such actions? Plan adequate time for discussion of these issues and opportunities for youngsters to share their feelings in a journal or through various "public relations" efforts (see previous activities).

Incredible Facts

♻ **Americans consume about 450 billion gallons of water every day.**
  • *How much water does the average American consume in one day?*
  • *How much water does your family consume in a week?*
  • *How much water does your family waste in a week?*

♻ **About 2.1 million tons of used motor oil finds its way into our rivers and streams every year.**
  • *How does your family dispose of used motor oil?*
  • *What happens to motor oil when it's poured on the ground?*
  • *How does motor oil affect the lives of freshwater creatures?*

♻ **A gallon of gasoline can pollute 750,000 gallons of water.**
  • *How long would it take your family to drink 750,000 gallons of water?*
  • *How much water do you drink in a year?*

Related Literature

Fredericks, Anthony D. *Simple Nature Experiments with Everyday Materials.* (New York: Sterling Publishing, 1995).
Nearly 100 activities and experiments help youngsters learn more about their natural world.

Hoff, Mary. *Our Endangered Planet: Groundwater.* (Minneapolis, MN: Lerner, 1991).
The importance of groundwater sources and how they are being used and polluted.

Hoff, Mary. *Our Endangered Planet: Rivers and Lakes.* (Minneapolis, MN: Lerner, 1991).
The value of rivers and lakes and how humans have polluted these bodies of water over the years.

Javna, John. *50 Simple Things Kids Can Do to Save the Earth.* (Kansas City, MO: Andrews and McMeel, 1990).
A guide to help kids take an active role in preserving their environment now and in the future.

Stille, Darlene. *Water Pollution.* (Chicago: Children's Press, 1990).
Animal, plant, and human needs for clean water are explored in this informative book.

Part IV

Dinosaurs— Jurassic Park

Chapter 11

Dinosaur Questions

Bernard Most

SAN DIEGO: HARCOURT BRACE, 1995

Book Summary

This book poses a series of questions that kids frequently ask about dinosaurs. Included are queries such as: "What did dinosaurs eat?"; "How smart were dinosaurs?"; "Which dinosaur ran the fastest?"; "Did dinosaurs swim?" Answers are clearly presented, often with a touch of humor.

Questions to Share

1. What are some questions you have about dinosaurs that were not presented in the book?
2. What are some resources or other reference material you could use to locate the answers to your questions?
3. What did you think was the most interesting question (or answer) presented in the book?

Major Book Project

A lot of what scientists know about dinosaurs has come from the reconstruction of dinosaur skeletons. Unfortunately, since no human being has ever seen a live dinosaur, the reconstruction of these skeletons is often based on speculation, some intelligent guesswork, and comparisons with present-day animals that may share some of the same features or characteristics.

The following activity demonstrates to children some of the difficulties scientists face when attempting to reconstruct a dinosaur skeleton (or a portion of a skeleton).

Place a whole (unboned) chicken in a pot of water to which has been added one or two cups of vinegar. Boil the chicken completely (until the meat almost falls off the skeleton). Remove as much meat as possible and continue to boil until all the meat can be removed with your hands (*Careful: The meat will be hot!*) Allow the chicken skeleton to cool completely and then carefully separate the individual bones. Allow the bones to dry in the open air for about two to three days.

Provide children with the pile of bones and invite them to reconstruct the entire chicken skeleton according to what they know a chicken looks like.

(*Note:* Children will discover many difficulties in reassembling the chicken skeleton even though they may have seen live chickens or illustrations of chickens. Inform youngsters that even though they may be familiar with chickens they may still encounter some difficulties in assembling a chicken skeleton. Invite them to imagine the difficulties of scientists reassembling a dinosaur skeleton [or part of one] when nobody has ever seen a live dinosaur.)

Special Project

Invite children to visit the school or public library and assemble a collection of dinosaur books for various age groups. For example, which books would be most useful for primary level (grades 1–3) readers; which ones for intermediate (grades 4–6) readers; and which one for readers at higher grade levels? Which books provide the most useful information or answer the questions kids most ask about dinosaurs. Youngsters may wish to establish their own evaluation system to rate and categorize these books. The following is a suggested format that can be duplicated and given to children for use as an evaluation guide:

	High				Low
1. The information is accurate	5	4	3	2	1
2. The information is up-to-date	5	4	3	2	1
3. The author is qualified to write this book	5	4	3	2	1
4. The organization is easy to understand	5	4	3	2	1
5. The illustrations/photos are good	5	4	3	2	1
6. The material is easy to understand	5	4	3	2	1

Field Trip

If you live near a college or university, call the biology department, zoology department, or archeology department and ask to speak to a "dinosaur expert." Set up an interview with a professor and the children. Invite children to assemble a list of some of their favorite questions prior to the visit. Most scientists are eager to share their information with youngsters and can offer some important up-to-date information concerning recent discoveries. The institution may also have some fossils or bones to share with youngsters.

Write Away

Several commercial companies sell authentic fossils or replicas of fossils. Invite youngsters to write to several of these companies and ask for a current catalog. When the catalogs arrive, invite students to read them and note the types of materials offered by each company. If possible, you may wish to order one or more fossils or dinosaurs eggs from these companies. Here are a few to get you started (listings of others can be found in the back pages of *Earth* magazine).

Phoenix Fossils
6401 East Camino De Los Ranchos
Scottsdale, AZ 85254

Older Than Dirt
Box 371
Moorpark, CA 93020

Natural History Supply House
12419 Coronet Drive
Sun City West, AZ 85375

Most dinosaurs were the size of chickens.
- *What are the dimensions of an average-sized chicken?*
- *How do the dimensions of a chicken compare with the dimensions of a Tyrannosaurus Rex?*

The tongue of an Apatosaurus weighed four tons.
- *How much does a human tongue weigh?*
- *How much does the tongue of an elephant weigh?*
- *Which present-day animal has the heaviest tongue?*

Ultrasaurus was able to raise its head to a height of 55 feet above the ground (that's taller than a five-story building).
- *How high is your house?*
- *How much taller than your house was an Ultrasaurus?*
- *How much taller than you was an Ultrasaurus?*

Related Literature

Aliki. *Fossils Tell of Long Ago*. (New York: HarperCollins, 1990).
A great introduction to the formation and creation of fossils highlighted by easy-to-understand illustrations.

Barton, Byron. *Bones, Bones, Dinosaur Bones*. (New York: HarperCollins, 1990).
This ideal book for young readers offers a glimpse into the search for dinosaur bones as well as how dinosaur skeletons are constructed.

Craig, Jean. *Discovering Prehistoric Animals*. (Mahwah, NJ: Troll, 1989).
A simple, straightforward text filled with lots of down-to-earth facts about dinosaurs.

Lasky, Kathryn. *Dinosaur Dig*. (New York: Morrow, 1990).
Several families are involved in the search for dinosaur fossils in this description of a dig in Montana.

Sattler, Helen. *The New Illustrated Dinosaur Dictionary*. (New York: Lothrop, 1990).
It's all here! Everything any dinosaur nut would want to know about 350 dinosaurs and other related creatures.

Chapter 12

Dinosaurs: Strange and Wonderful

Lawrence Pringle

HONESDALE, PA: BOYDS MILLS, 1995

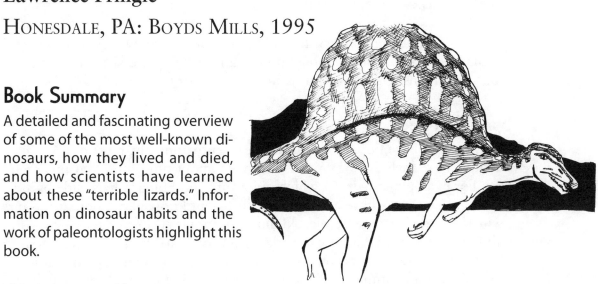

Book Summary

A detailed and fascinating overview of some of the most well-known dinosaurs, how they lived and died, and how scientists have learned about these "terrible lizards." Information on dinosaur habits and the work of paleontologists highlight this book.

Questions to Share

1. What is the most interesting fact you know about dinosaurs?
2. What would you think would be the most interesting part of a paleontologist's job?
3. Which dinosaur is your favorite? What habits or physical characteristics make it your favorite?
4. What do you think is the greatest mystery surrounding dinosaurs? How do you think scientists will solve that mystery?

Major Book Project

Dinosaurs came in all shapes and sizes. Children are sometimes amazed at the height, weight, and length of some of the more popular dinosaurs. However, in order to fully appreciate dinosaurs sizes they need a frame of reference or a comparison feature. Here's one way to help youngsters appreciate these gigantic sizes.

Measure various lengths of string, twine, or yarn according to the measurements on the next page. Go to a backyard, driveway, or school playground and lay out several pieces of string side by side. Invite children to mark the ends of each piece of string with chalk. Encourage youngsters to compare the distances by lying down next to them, walking them off with their feet, or some other comparable form of measurement.

Dinosaur	Length
Compsognathus	2 feet
Velociraptor	8 feet
Stegosaurus	28 feet
Triceratops	30 feet
Tyrannosaurus Rex	32 feet
Brachiosaurus	67 feet
Ultrasaurus	100 feet

Afterward, invite children to make a list of objects with which they are familiar (school bus, automobile, house) and develop a chart that compares dinosaur lengths with objects in youngsters' environment.

Special Project

Many movies have popularized dinosaurs. Unfortunately, most of those productions have included or emphasized facts about dinosaurs that just aren't true. For example, movies in which dinosaurs and humans lived side by side.

After youngsters have accumulated sufficient knowledge about dinosaurs invite them to view several selected dinosaur movies (most larger video stores will have these available or can order them for you). Encourage children to observe the misrepresentations or erroneous information presented in some of these films (for example, some scenes in the movie *Jurassic Park,* which was one of the most authentic movies about dinosaurs, did stretch the truth for dramatic effect). Encourage youngsters to assemble an informational chart that lists selected "untruths" discovered in various videos. The chart should also contain corrections for the erroneous data.

Write Away

Invite children to put together a "Dinosaur Newspaper" that presents interesting facts and observations about dinosaurs in a newspaper format. Invite them to use the same sections as the local newspaper (e.g., sports—how fast some dinosaurs were able to run; fashion—speculation on the colors of various dinosaurs; food and health—the different types of diets of meat-eaters and plant-eaters). The newspaper can be assembled using a word processing program and printed for distribution to family and friends.

Arts and Crafts

Different artists have illustrated, drawn, and painted dinosaurs in different ways throughout the years. For example, the most famous dinosaur of all—Tyrannosaurus Rex—has been depicted by various artists as slow and dumb or as cunning and fast.

Invite children to assemble a collection of drawings and illustrations of selected dinosaurs for display. They may wish to search through encyclopedias, different trade books on dinosaurs, or scientific magazines for different renditions of identical dinosaurs. Invite children to discuss some of the reasons why different artists have depicted the same dinosaur in various ways. What are some of the similarities? What are some of the differences?

🦕 **One of the largest known dinosaurs was Ultrasaurus, which was fifteen times larger than an African elephant.**
- *How large is an African elephant?*
- *How much larger is the Ultrasaurus in comparison with a family pet (dog, cat, etc.)?*
- *How does the Ultrasaurus compare with a standard automobile?*

🦕 **Many scientists believe that the closest living relatives of dinosaurs are birds.**
- *What are some of the similarities between dinosaurs and birds?*
- *How are the feathers of a bird similar to the scales of a dinosaur?*

🦕 **The Stegosaurus, which weighed up to 2 tons, had a brain that weighed only $2\frac{1}{2}$ ounces.**
- *How much does an average human brain weigh?*
- *How much heavier is your brain than the brain of a Stegosaurus?*
- *What can you do with your brain that a Stegosaurus couldn't do with its brain?*

Related Literature

Cosner, Shaaron. *Dinosaur Dinners*. (New York: Watts, 1991).
This book concentrates on the food that dinosaurs ate and offers additional information on the history and evolution of dinosaurs.

Hisa, Kunihiko. *How Did Dinosaurs Live?* (Minneapolis, MN: Lerner, 1990).
This book examines the lifestyles of 37 different types of dinosaurs and how they lived.

Lambert, David. *Dinosaurs*. (New York: Warwick, 1989).
A well-written book that presents general information about dinosaurs. It also includes some activities of various dinosaurs.

Schlein, Miriam. *Discovering Dinosaur Babies*. (New York: Four Winds, 1991).
Lots of information about the ways in which different species of dinosaurs cared for their young.

Wexo, John. *Dinosaurs*. (Mankato, MN: Creative Educations, 1991).
An informative book highlighted by interesting and creative illustrations throughout. Lots of information here.

The Meat Eaters Arrive

Suzan Reid

BUFFALO, NY: FIREFLY BOOKS, 1996

Book Summary

The Brontos lived in a very quiet neighborhood planting and eating shrubs and trees. Then, the Rexes moved in next door and the Brontos feared for their lives. A humorous depiction of plant-eating and meat-eating dinosaurs in a suburban environment.

Questions to Share

1. What dinosaur features were accurately portrayed through the illustrations?
2. How did this book differ from most dinosaur books?
3. What modern-day elements were included in this story to make it humorous?

Major Book Project

Much of what scientists know about dinosaurs is revealed by the fossils of those creatures. Through fossils, paleontologists can determine whether a dinosaur was a plant-eater or a meat-eater. Other features such as size and weight, bone structure, speed, and habits can be determined through an examination of fossils.

The following activity allows kids to create their own fossils and to examine them in much the same way as scientists out in the field.

Fill an aluminum pie pan with wet sand (not too wet, not too dry). Invite children to press the skeleton of a fish or several chicken bones into the sand to create an impression. Carefully remove the bones (tweezers may be necessary). Invite children to fasten the ends of a cardboard strip (approximately 1 inch high by 10 inches long) together to form a circle. Have them place the strip into the sand around the bones' impression (being careful not to disturb the impression). Mix up some plaster of paris (available at any craft or variety store) according to the package directions. Children can pour the mixture carefully into the cardboard ring. Allow it to dry for about one hour. Remove the cardboard ring and lift the plaster from the plate. Kids can brush off any excess sand and observe their "fossil."

Invite youngsters to record their observations of the fossil's details in a journal or notebook. What features tend to stand out? Which ones are difficult to observe? Later, children may wish to place their fossil outside (to expose it to the elements) or bury it underground and after several months note how much it has deteriorated. Comparisons between the "original" fossil and its deteriorated condition can be made.

Special Project

Encourage youngsters to take photographs or draw illustrations of animal jaws with which they are familiar (i.e., dog, cat, hamster, cow, etc.). What similarities do they note in the jaws of these modern-day animals? Are there any similarities between the jaw and teeth of a modern-day plant-eater (such as a cow) and the jaw and teeth of a plant-eating dinosaur (such as a Saltasaurus)? Are there any similarities between the jaw and teeth of a modern-day meat-eater (such as a dog) and the jaw and teeth of a meat-eating dinosaur (such as Tyrannosaurus Rex)? Invite children to create charts and graphs of the similarities and differences (based on diet) between selected modern-day animals and specific dinosaurs.

Creative Dramatics

Invite children to create a short vignette or skit depicting a meeting between a meat-eating dinosaur and a plant-eating dinosaur. Using information gathered from several references, encourage children to compose a chance encounter between these two different types of dinosaurs. Would they be friendly? How much would size and weight make a difference in the encounter? If they were both the same size, would they ignore each other? Children may elect to present this brief dramatic presentation to another group of individuals or videotape it for sharing with others.

Across the Curriculum

Language Arts

Invite children to locate books in the library about dinosaurs—particularly those that refer to plant-eaters and meat-eaters. Invite them to gather information and record that data in a chart similar to the one below:

	Meat-eaters	**Plant-eaters**
Basic Facts		
Species		

Invite youngsters to contact paleontologists or other dinosaur experts at a nearby college or university to inquire about the skull and jaw features of plant-eating dinosaurs and meat-eating dinosaurs. That data can also be recorded in the chart.

Geography

Dinosaur remains have been discovered in many states throughout the United States. Provide youngsters with a list of those states (following) and encourage them to locate information in trade books or library references listing representative examples of uncovered dinosaurs in each state. Children can post illustrations of those dinosaurs around a wall map of the United States with a length of yarn tacked to the picture at one end and to the indicated state at the other end.

States where dinosaurs have been discovered: Alabama, Alaska, Arizona, Arkansas, Colorado, Connecticut, Kansas, Maryland, Massachusetts, Michigan, Mississippi, Missouri, Montana, New Jersey, New Mexico, North Carolina, Oklahoma, South Dakota, Texas, Utah, and Wyoming.

Incredible Facts

The biggest dinosaur teeth discovered belonged to Tyrannosaurus Rex. They were more than 6 inches long.
- *How long are your teeth?*
- *How much longer are T. Rex's teeth than yours?*
- *What modern day animal has the longest teeth?*

The claw of Therizinosaurus, a dinosaur of the Cretaceous period, was $27\frac{1}{2}$ inches long.
- *How large is your hand?*
- *How does the size of your hand compare with the claw of Therizinosaurus?*
- *What are some modern-day clawed animals?*

Corythosaurus, a plant-eating dinosaur, had hundreds of grinding teeth packed together in its jaw.
- *How many teeth do you have?*
- *What are the different types of teeth found in a human jaw?*
- *Which of your teeth are used for eating meat? Which are used for eating plants?*

Related Literature

Dixon, Dougal. *Dougal Dixon's Dinosaurs.* **(Honesdale, PA: Boyds Mills, 1993).**
An exciting, up-to-the-minute compilation of everything scientists know about dinosaurs. This
    is one of the best reference guides around.

Hopkins, Lee Bennett. *Dinosaurs.* **(San Diego: Harcourt Brace, 1987).**
Eighteen poems give youngsters some fresh perspectives and delightful insights into the world
    of dinosaurs.

Prelutsky, Jack. *Tyrannosaurus Was a Beast.* **(New York: Greenwillow, 1988).**
You won't want to miss this delightful collection of dinosaur poems. A wonderful array of
    humorous verse.

Sattler, Helen. *Tyrannosaurus Rex and Its Kin: The Mesozoic Monsters.* **(New York: Lothrop,
    1989).**
The book examines the most famous of all dinosaurs and takes a look at its descendants.
    Colorful illustrations and a time line highlight this book.

Simon, Seymour. *New Questions and Answers about Dinosaurs.* **(New York: Morrow, 1990).**
Scientists are learning more about dinosaurs every day. This book presents readers with the
    most up-to-date information.

Part V

Earth—
Can You Dig It?

Chapter 14

In Coal Country

Judith Hendershot

New York: Knopf, 1987

Book Summary

A young girl grows up in a coal-mining town in Ohio. The trials and tribulations of family life are wonderfully told in this personal tale, which weaves narration with factual data. The excitement and hard work that make up this lifestyle is realistically portrayed through the eyes of the narrator. Magnificent illustrations highlight the story.

Questions to Share

1. What would you find most interesting about living in a coal-mining town?
2. What parts of the story were most like your own childhood? What similarities do you have with the main character?
3. Should everyone be proud of the work they do no matter how difficult or dirty it is?
4. What questions would you like to ask the author of this book?

Major Book Project

Provide children with one hard chocolate chip cookie, one soft chocolate chip cookie, and some toothpicks. Tell them that the chocolate chip cookies simulate coal mining areas and that the chips are the pieces of coal that they will be mining from those areas. Before beginning, encourage youngsters to guess which cookie will be easier to "mine"—the hard one (which represents anthracite, or "hard" coal) or the soft one (which represents bituminous, or "soft" coal).

Invite youngsters to mine the chips from each cookie using the toothpicks as their only tools. Encourage them to obtain the chips with a minimum amount of "destruction" to each cookie surface. Give them a designated period of time (e.g., four minutes per cookie) for their "mining" operation.

Afterward, invite children to make comparisons between the "mining" of chips on each of the two cookies and the mining of coal in the real world. (*Note:* Children will realize that it is easier to "mine" chips from the hard cookie than it is from the soft cookie. This is also true in the coal-mining industry—it is easier to mine anthracite coal [there is less destruction to the environment] than it is to mine bituminous coal.)

Engage children in a discussion of how humans can protect the Earth while providing for the energy needs of the people who live on the Earth.

Special Projects

A. An excellent video on coal mining is *Portrait of a Coal Miner* produced by the National Geographic Society (Washington, D.C.; catalog no. 51175), which portrays the life of a coal-mining family, how coal is mined, and what life is like inside and outside a coal mine. If possible, obtain a copy of this video and show it to children. Afterward, invite them to compare the video with the book and with their own lives. What similarities and/or differences did they note? Children may want to complete a chart similar to the one below:

	In Coal Country	**Portrait of a Coal Miner**	**My Life**
Work			
Play			
Family			

B. Invite students to complete some of the "sentence stems" below:

"If my father was a coal miner, I would … ."
"If my family did not have a television, we would … ."
"If the rivers near us were black, I would … ."
"If I only took a bath once a week, I would … ."
"If I lived in a coal town, I would … ."

Field Trip

Invite students to look through the local yellow pages of the telephone book to determine if coal is sold in your community. If it is, make arrangements for children to visit a coal distributor. They may wish to assemble a list of questions to ask the distributor. Queries may include the following:

Where does the coal come from?
How is it mined?
Who buys the coal?
What is the coal use for?
How much coal is used by a typical family?

Encourage youngsters to record the responses in a journal or notebook for future reference.

If coal is not sold in your community, obtain copies of telephone books from metropolitan areas throughout the country. These are usually available at most large public libraries. Borrow one or more and invite students to conduct a search through the yellow pages for coal suppliers and distributors. Discuss with kids reasons why coal may be sold more in one part of the country than in another part. Talk about reasons why it may be difficult to locate a supplier of coal in your area or in another area of the country. What factors contribute to the availability or distribution of coal in one region versus another region?

Experiment

One of the distinctive features about living in a coal-mining town is the amount of coal dust that is in the air, and that it settles on everything and everybody. While the children with whom you work may not live in a coal-mining area or ever intend to live in a coal-mining area, they can get some sense of air pollution with the following experiment.

Invite children to take 3 to 4 index cards and smear a thin layer of petroleum jelly on one side of each card. Encourage youngsters to place the cards in various locations in and around where they live (i.e., in a tree, taped to the side of the house, and taped to a pencil stuck in the ground). The cards should be placed at varying heights around the house.

After a designated period of time (i.e., one week, three weeks, six weeks) encourage children to remove the cards and note the amount of particulate matter (air pollution) that has settled on the surface of each card. In which location was the pollution the greatest? What would be the effects of inhaling that pollution over the same length of time? Help youngsters draw relationships between their discoveries and the everpresent coal dust described in the book.

🌐 **Some coal deposits may be more than 300 million years old.**
- *Where are some of the largest deposits of coal in the United States?*
- *Where are some of the largest deposits of coal in the world?*

🌐 **Coal is a fossil fuel made from the remains of plant materials compressed into rock.**
- *What are some other fossil fuels?*
- *Where are some large deposits of fossil fuels?*
- *Which fossil fuel is used most often in the world?*

Related Literature

Bramwell, Martin. *Understanding and Collecting Rocks and Fossils.* (Saffron Hill, England: Usborne Publishing, 1983).
A well-written and comprehensive guide to the rocks and fossils of the world and how they were created.

Farndon, John. *How the Earth Works: 100 Ways Parents and Kids Can Share the Secrets of the Earth.* (New York: Reader's Digest, 1992).
Lots and lots of experiments and projects designed to help kids learn about the Earth.

Fichter, George. *Rocks and Minerals.* (New York: Random House, 1982).
Descriptions of more than 75 rocks and minerals as well as methods for locating them are included.

Loeschnig, Louis. *Simple Earth Science Experiments with Everyday Materials.* (New York: Sterling Publishing, 1996).
Dozens of explorations, discoveries, and experiments designed to help youngsters learn more about their planet.

Parker, Steve. *Rocks and Minerals.* (New York: Dorling Kindersley, 1993).
Lots of information and lots of activities for collecting a wide variety of rocks and minerals.

Chapter 15

Mojave

Diane Siebert

NEW YORK:
HARPERCOLLINS, 1988

Book Summary

A beautiful, lyrical poem about life on one of the world's most diverse ecosystems—the desert. In particular, this book describes the flora and fauna of the Mojave Desert in southern California, the wonderful interplay of plants and animals and how they survive, the diversity of relationships in and around the desert, and the incredible variety of goings-on that make the desert such a rich environment.

Questions to Share

1. What surprised you most about this book?
2. If you could be one of the creatures in this book, which one would you be?
3. How is the desert similar to or different from the place in which you live?
4. How did the illustrations contribute to your appreciation of this story?

Major Book Project

Children can create their own miniature desert environment in their own homes. A desert terrarium is a small controlled environment containing a variety of plants and/or animals in an artificial situation that closely imitates the natural living conditions of those organisms. A desert terrarium can endure for a long period of time and provide youngsters with a close-up look at a "sample" of nature. Share the following directions with the children with whom you work:

1. Fill the bottom of a large glass container (an old aquarium purchased at a pet store or garage sale, a large pickle jar, or a 2-liter plastic soda bottle) with a layer of coarse sand or gravel. Combine one part fine sand with two parts potting soil and spread this over the top of the first layer.
2. Sprinkle this mixture lightly with water.

3. Place several varieties of cacti (small potted varieties are available at most gardening centers or nurseries) into the terrarium (wear gloves to avoid being pricked by the cacti).
4. When planting the cacti, be sure that the roots are covered completely by the sandy mixture.
5. Children may wish to place several desert animals such as lizards and horned toads in the terrarium. Be sure the animals have a sufficient quantity of food and water available as well as some type of dwelling (a small can, a few rocks, etc.). Inquire at a local pet store about the availability of selected pets as well as the care and feeding of those animals.
6. The terrarium can be left in the sun, although it should be lightly sprinkled with water about once a week. For the cover of the terrarium cut a section out of an old nylon stocking and secure it to the top with a rubber band.

Invite children to maintain a journal or diary of life inside the terrarium. What changes take place in the organisms? How do they react to each other? How do they grow or move around in their miniature environment? Plan time to talk with youngsters about these events.

Special Project

Visit a local gardening center or nursery with children and investigate the variety of cacti for sale. If possible, select several varieties and take them home. Invite youngsters to become guardians of these plants. They can create special journals or logs for their "adopted" plants. Provide children with examples of baby books and invite them to create a "cactus book" based on the format of typical baby books. Included could be a section on the "baby's" first year, a section of first photographs, significant measurements (height, width, weight), rate of growth, and other significant details. Invite children to share their information with others on a regular basis.

Creative Dramatics

The book begins and ends with the following lines:

> I am the desert.
> I am free.
> Come walk the sweeping face of me.

Invite children to dramatize those lines. How would they perform those lines for other people? How could they bring those lines to "life" in a dramatic presentation? Afterward, encourage children to develop similar lines for places in their environment and to act out those lines for others. Following are some possible examples:

> I am the river.
> I am free.
> Come swim the shimmering blue of me.

I am the playground.
I am free.
Come run the painted lines of me.

I am the yard.
I am free.
Come dance among the trees of me.

Write Away

Children may wish to obtain additional information about the Mojave Desert. Have them write to the following places and obtain relevant brochures and information sheets. When the materials arrive, invite students to create a descriptive display of life in this special ecosystem. Using the information in the book along with that obtained through the mail, youngsters can assemble a poster, bulletin board, booklet, or some other type of visual display about the Mojave.

California Desert Information Center
831 Barstow Road
Barstow, CA 92311

Death Valley National Park
P.O. Box 579
Death Valley, CA 92328

Joshua Tree National Park
74485 National Park Drive
Twentynine Palms, CA 92277

Across the Curriculum

Geography, Language Arts

In the United States there are four principle desert regions. These include the Mojave, the Sonoran, the Chihuahuan, and the Great Basin Deserts. Many youngsters believe that deserts are all the same and that the same types of plants and animals live in every desert region. In fact, there is a wide variety of differences among the deserts of the world as there is among the deserts of the United States. While the climatic conditions are often similar, the flora and fauna can be vastly different.

Provide youngsters with an opportunity to read additional books about U.S. deserts (see the listing on the next page). Invite them to gather information about plant and animal life in those different regions. Encourage them to create and complete a chart such as the following:

	Animals	Plants
Mojave		
Sonoran		
Chihuahuan		
Great Basin		

After the chart has been completed allow children an opportunity to discuss the similarities and/or differences between the four major deserts of the United States.

A saguaro cactus can weigh as much as a full-grown elephant.
- *How much does an elephant weigh?*
- *What is the heaviest plant in your house?*
- *How tall does a saguaro cactus grow?*

The surface of the desert can have a temperature of 170°F or more.
- *What is the hottest temperature recorded in your town?*
- *What is the difference between the temperature today and the temperature of the desert floor?*
- *How hot does an oven need to be to bake cookies?*

About one-half of Australia is desert.
- *How large is Australia?*
- *What are some of the plants and animals that live in Australian deserts?*
- *How are those organisms similar to those that live in U.S. deserts?*

The major portion of the world's deserts are not sand dunes but rather rocky or mountainous terrain.
- *Where are some of the major deserts in the world?*
- *How does the temperature in those deserts compare with the average temperature in U.S. deserts?*
- *What is the largest desert in the world? What is the smallest?*

Related Literature

Dewey, Jennifer. *A Night and Day in the Desert.* **(Boston: Little, Brown, 1991).**
A fascinating collection of vivid word pictures of action taking place in the desert. Great illustrations.

Silver, Donald. *Cactus Desert.* **(New York: W. H. Freeman and Company, 1995).**
An amazing and incredible journey through this remarkable ecosystem. Lots of things to do and lots to discover.

Stephen, Richard. *Deserts.* **(Mahwah, NJ: Troll, 1990).**
An excellent reference of the plants, animals, and people who live in this harsh environment.

Twist, Clint. *Deserts.* **(New York: Dillon Press, 1991).**
Desert ecology is examined with an emphasis on the abundant life found in the desert.

Wallace, Marianne. *America's Deserts.* **(Golden, CO: Fulcrum Publishing, 1996).**
A descriptive and heavily illustrated examination of desert life. A wonderful colorful array of plants and animals.

Chapter 16

The Village of Round and Square Houses

Ann Grifalconi

BOSTON: LITTLE, BROWN, 1986

Book Summary

In the Cameroons of central Africa exists an isolated village named Tos. In that village the women live in round houses and the men live in square houses. The story of how this came to be is told through the eyes of a young girl as she shares a beautiful legend about a community and its people.

Questions to Share

1. How is the village of Tos similar to the town or city where you live?
2. What did you enjoy most about life in the village of Tos (before the volcanic eruption)?
3. How is the narrator similar to you, a family member, or one of your friends?
4. How do natural events, such as erupting volcanoes and earthquakes, change the way people live?
5. How would your life change if a volcano erupted nearby?

Major Book Project

Children may wish to create their own chemical volcano. They can participate in the gathering and setting up of materials. This activity, however, should be done under adult supervision only.

Set an empty soda bottle on the ground or in the sand of a playground sandbox. Mound up the dirt or sand around the bottle so that only the top of the neck shows. Pour one tablespoon of liquid detergent into the bottle. Add a few drops of red food coloring, one cup of vinegar, and enough warm water to fill the bottle almost to the top. Very quickly, add two tablespoons of baking soda to the bottle. (The baking soda can be mixed with a little water before-hand. This will make it easier to pour it into the bottle without spilling.)

Children will note that the artificial volcano "erupts" in a similar way to many volcanoes. In order to provide children with a frame of reference to which they can compare their "homemade" volcano, obtain a copy of either of the following two videos which are available from the National Geographic Society (Washington, D.C.): *This Changing Planet* (catalog no. 30352) or *The Violent Earth* (catalog no. 51234). Provide opportunities for youngsters to share the similarities and differences between the volcanoes depicted in the films and their personal volcano.

Special Project

Invite children to complete a chart similar to the one below. Ask them to note any similarities in the customs, traditions, or behaviors of men and women in their town or city in comparison with the customs of men and women in the village of Tos. Provide opportunities for youngsters to discuss those comparisons.

	Customs—Men	Customs—Women
Tos		
My Town		

Creative Dramatics

Invite children to create a brief five-minute skit or a short one-act play about the village and villagers both before and after the eruption of the volcano. Encourage youngsters to consider the roles and occupations of both men and women before the eruption and how those roles may have changed after the eruption. What customs or traditions may have remained the same? Which ones might have changed as a result of the men living in one type of house and the women in another? Children may wish to create a short dialogue between two or three characters (men and women) that would have taken place before the eruption and the same characters in an altered dialogue after the eruption. Plan time to discuss any changes in behavior, attitudes, or relationships that may or may not have existed between men and women after the eruption.

Write Away

Invite children to write to several of the following science supply houses and inquire about the availability of volcanic ash. They also may wish to contact the earth sciences department at a local college or university to obtain a small sample of ash.

Central Scientific Company
3300 CENCO Parkway
Franklin Park, IL 60131-1364
(800) 262-3626

Delta Education
P.O. Box 3000
Hudson, NH 03061-3000
(800) 282-9560

Frey Scientific
905 Hickory Lane
Mansfield, OH 44905
(800) 225-FREY

Across the Curriculum

Language Arts, Geography

Provide youngsters with copies of different newspapers from selected cities around the country. (Many metropolitan areas have newsstands at which different newspapers from various cities are sold.) If it is not possible to obtain a variety of newspapers, the hometown paper will suffice. Invite youngsters to look through those newspapers for articles, information, or data relating to volcanic eruptions. (*Note:* At the time of this writing, Mount Kilauea volcano in Hawaii is in a constant state of eruption. Occasional articles about the process periodically appear in the local newspapers.) Invite children to cut out those articles and assemble them into an on-going journal. Index cards with a brief summary of the date, location, and events surrounding an eruption can be posted around a large wall map for others to read.

Reading

Obtain a copy of the following book from your school or local public library: *Volcanoes* by Seymour Simon (New York: Mulberry, 1988). Provide opportunities for children to read this book. Afterward, invite youngsters to compare the photographs in the book with volcano photos in other books. Encourage children to categorize the photos from various sources according to one of the four types of volcanoes—shield volcanoes, cindercone volcanoes, composite volcanoes, and dome volcanoes. Invite children to categorize the volcano in *The Village of Round and Square Houses* into one of these four types.

Experiment

In the book, the villagers discovered that their soil was improved after the eruption of the volcano. The soil was richer and crops flourished.

Encourage youngsters to mix different amounts of ash with equal amounts of potting soil using the following ratios (1:1, 1:2, 1:3, and 1:4). Invite them to fill several compartments of an egg carton with the different mixtures and label each compartment. Encourage them to select two or more different varieties of plant seeds (radish, bean, corn) and plant several vegetable seeds in each of the different compartments. Encourage youngsters to make predictions on the soil/ash mixture that will be most conducive to seed germination. Have youngsters compare the relative growth rates of the vegetables. In which growth medium do the seeds germinate first? Which one is most conducive to healthy growth? How does the amount of volcanic ash affect the germination and growth of selected plants?

The planet Mars has an enormous volcano—called Olympus Mons—which is 16 miles high and 370 miles wide.
- *What is the tallest volcano in the world?*
- *What is the widest volcano in the world?*
- *What is the tallest volcano in the United States?*

Scientists consider a volcano to be active if it has erupted anytime in the last ten thousand years.
- *Which country has the most active volcanoes?*
- *Which U.S. state has the most active volcanoes?*
- *Why are most volcanoes located around the rim of the Pacific Ocean?*

The temperature of lava can be more than 2000°F.
- *How hot is boiling water?*
- *How much hotter is lava than boiling water?*
- *What are some of the destructive effects of lava?*

In 1883 the volcanic island of Krakatoa exploded with such a loud bang that the sound was heard 2,500 miles away.
- *Are all volcanoes noisy?*
- *What have been some of the noisiest eruptions in the world?*

Related Literature

Lasky, Kathryn. *Surtsey: The Newest Place on Earth.* (New York: Hyperion, 1992).
Lyrical prose and spectacular photographs recount the story of this newest volcanic island created off the coast of Iceland in 1963.

Lauber, Patricia. *Volcano: The Eruption and Healing of Mount St. Helens.* (New York: Bradbury, 1986).
A documentation and explanation of one of the most incredible eruptions of modern times.

Simon, Seymour. *Volcanoes.* (New York: Mulberry, 1988).
The photographs are unbelievable and the text is clear and direct. A most engaging book.

Taylor, Barbara. *Mountains and Volcanoes.* (New York: Kingfisher, 1993).
A wonderfully illustrated text highlighted by informative details and stories about mountains and volcanoes.

Watt, Fiona. *Earthquakes and Volcanoes.* (London, England: Usborne, 1993).
Lots of illustrations and detailed text make this a fascinating book on volcanoes and earthquakes.

Chapter 17

?What If ... The Earth

Steve Parker

BROOKFIELD, CT:

COPPER BEECH BOOKS, 1995

Book Summary

An engaging book that shares important Earth–related information with readers through a series of "what if" questions. Included are queries such as: "What if the Earth stood still?"; "What if the continents didn't move?"; "What if there were a lot more volcanoes?"; "What if there was no more soil?" An ideal introduction to earth science.

Questions to Share

1. What "what if" questions about the Earth do you have that were not answered in this book?
2. Which of the questions in the book would you like to explore further?
3. How did the illustrations contribute to your understanding of the book?

Major Book Project

The soil in our backyards or gardens is composed of many different materials or elements. By digging into the soil we may discover a host of various products. The many different materials that make up soil form layers depending on their composition or relative weight. In this project various soil samples are mixed with water to demonstrate how sediment layers are formed.

1. Obtain different soil samples from the neighborhood or surrounding areas. Each sample should include at least one cup of soil. The samples can include humus, garden soil, peat moss, and clay.

2. For each sample of soil, provide children with a jar and a watertight screw-on lid (baby food jars work particularly well). Ask children to place approximately one or two ounces of each soil sample in a different jar. Invite children to examine the components of each soil sample and record its unique characteristics.
3. Invite children to fill each jar approximately half-full with water and to put the lid on. Instruct kids to shake each jar for 30 seconds.
4. After shaking, ask children to observe the soil and water mixture in each jar and record their observations. Ask children to leave the different jars standing for several days and to observe what happens inside each jar. Invite children to draw illustrations of each soil sample and the layers that formed inside each jar.

Soil is composed of numerous particles, ranging from minute sandlike pieces to larger pieces of gravel or wood. When soil is mixed with water and allowed to settle, layers are formed. The bottom layer contains the largest and heaviest particles, with successively higher layers containing smaller and smaller particles. The lightest particles float on the surface of the water.

Special Project

Invite children to take on the role of a particular type of rock (sedimentary, metamorphic, igneous). Ask them to create a diary of their selected rock's "life history."(Each child should assume the "identity" of a particular rock and write entries from the perspective of that rock.) They may wish to include information on the formation or creation of the rock, where it "lives," some of the events it "saw" during its "lifetime," and what it is doing now. Plan opportunities for children to share their diaries with others.

Write Away

Invite children to write to one or more of the following organizations. Each of these groups has a variety of newsletters, reports, brochures, and flyers about the Earth, with particular emphasis on how humans can help preserve the Earth's surface.

America the Beautiful Fund
219 Shoreham Building
Washington, D.C. 20005

The Izaak Walton League of America
1401 Wilson Boulevard, Level B
Arlington, VA 22209

National Recycling Coalition
1101 30th Street NW, Suite 305
Washington, D.C. 20007

Sierra Club
730 Polk Street
San Francisco, CA 94109

Across the Curriculum

Language Arts

A. Invite children to each construct a dictionary booklet entitled, "My Earth Book." Have youngsters make an original book consisting of at least 26 pages. When kids have completed constructing their books, have them each write one letter of the alphabet at the top of each page. Invite them to look through this book as well as other literature (see the related literature at the end of this section) for possible words. Encourage children to write about each word, draw pictures describing each word, or cut out pictures from old magazines that correspond to the word.

B. Encourage children to look through the daily newspaper for articles regarding any changes in the Earth's surface (e.g., earthquakes, floods, hurricanes, severe erosion, and volcanoes). Invite them to create a miniature bulletin board on which they can display the articles under the heading "The Changing Earth." Be sure to provide opportunities for children to discuss some of the articles with you as well as the implications of these events on the surface of the Earth.

Experiment

Obtain several samples of sandstone from a local building supply dealer or large hardware store. Invite children to soak the samples in water overnight. Then have children take each sample and place it inside a sealable plastic bag. Tell them that you will be placing this bag into the freezer overnight. Ask children to make predictions about what will happen to the sandstone inside each bag. Put the bags in the freezer and remove the next day. Ask children to observe the results. The sandstone will have cracked and broken apart because water, when it freezes, expands. Thus the water inside the sandstone expanded and cracked the rock. This process also occurs in nature when water seeps into the cracks in rocks, freezes, and cracks the rock more. This is how water is able to break down large rocks into smaller stones and pebbles.

Incredible Facts

🌐 **There are approximately 455 active volcanoes in the world.**
- *How many volcanoes erupt around the world each year?*
- *How many active volcanoes are in the United States?*
- *What was the most powerful volcanic eruption in history?*

(◑) Each year the United States loses 7 billion tons of topsoil.
- *How many pounds is that?*
- *How much topsoil is lost in your state each year?*
- *What causes the greatest loss of topsoil—floods or erosion?*

(◑) If the Earth was completely smooth, water would cover the entire globe to a depth of about 7,500 feet.
- *How many yards is that?*
- *What is the deepest point in the ocean?*
- *Why isn't the Earth completely smooth?*

(◑) The motion of the Earth's plates separates Australia and Hawaii by 2.7 inches each year.
- *How far apart are Australia and Hawaii right now?*
- *How far apart will Australia and Hawaii drift in one century?*
- *How were the islands of Hawaii created?*

Related Literature

Bain, Iain. *Mountains and Earth Movements.* (New York: Watts, 1984).
What are the effects of erosion, weathering, faulting, folding, and continental drift on the creation of mountains? This book describes all of them in detail and includes colorful photographs.

Bramwell, Martyn. *Understanding and Collecting Rocks and Fossils.* (London: Usborne, 1983).
Everything the young rock hound needs to identify, sort, and collect some of the more common rocks, fossils, and minerals.

Fichter, George. *Rocks and Minerals.* (New York: Random House, 1982).
A quick and easy reference guide to more than 75 different rocks and minerals.

Fodor, R. V. *Chiseling the Earth: How Erosion Shapes the Land.* (Springfield, NJ: Enslow, 1983).
The effects of erosion on the Earth's surface and the efforts being used to curb this natural phenomenon in various areas of the United States.

Hiscock, Bruce. *The Big Rock.* (New York: Atheneum, 1988).
The story of a single rock in the Adirondack Mountains and the weathering, erosion, and mountain-building that created it.

McGovern, Tom. *Album of Rocks and Minerals.* (Chicago: Rand, 1981).
This book is a marvelous introduction to geology and how various kinds of rocks were formed.

Nixon, Hershell and Joan Nixon. *Earthquakes: Nature in Motion.* (New York: Dodd, 1981).
This book is designed to answer all the questions students may have about earthquakes including how they occur, where they occur, how they are measured, and the amount of damage they cause.

Selsam, Millicent. *First Look at Rocks.* (New York: Walker, 1984).
For youngsters who need an easy-to-follow guide to rock collecting, this book fits the bill.

Srogi, LeeAnn. *Start Collecting Rocks and Minerals.* (Philadelphia, PA: Running Press, 1989).
This book describes the collection, identification, qualities, and location of rocks, minerals, and crystals.

Part VI

Oceans—
Wet and Wild

Chapter 18

One Small Square: Seashore

Donald M. Silver

NEW YORK: W. H. FREEMAN AND COMPANY, 1993

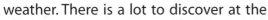

Book Summary

In this book, readers take a close look at one small section of the planet Earth—the seashore. They meet a dazzling array of creatures and watch how they interact with each other and with other elements of their environment including plants, rocks, soil, and the weather. There is a lot to discover at the seashore and this book offers young adventurers the information and guidance they need to make some magnificent discoveries. One of a series (see page 92).

Questions to Share

1. What was the most amazing animal you discovered by reading this book? What was the most amazing plant?
2. Would you like to explore the shoreline for animals and plants? Which one would you like to look for?
3. How is life on the seashore similar to life in your neighborhood, community, or town?
4. Where is the nearest seashore to where you live? How is it similar to the seashore(s) described in this book?

Major Book Project

Tide pools are areas of the seashore where pools of water have been left behind when the ocean recedes at low tide. These pools may be trapped in the nooks and crannies of rocks, between rock ledges and sand, or in depressions along the beach. Water in these pools is filled with a wide variety of plants and animals—some are microscopic, while others are larger and can be seen quite easily.

Tide pools can be interesting and exciting places for children to visit. Besides the wide variety of plants and animals there to discover, they will also learn how these organisms live where they do, and how they are able to survive from one day to the next.

But tide pools are fragile environments. If not careful, the human visitors who come to observe can damage or threaten this delicate ecosystem. Here are some suggestions to follow whenever you and the children with whom you work investigate a tide pool. Being aware of these points will help make the visit informative for humans and safe for the creatures you study.

- Remember that a tide pool is home for many creatures. Don't allow children to remove any organisms from their natural environment. The plants and animals will probably be unable to survive away from the seashore.
- Turn over any rocks carefully. Kids will discover lots of living creatures under the rocks. Afterward, return the rocks to their original positions.
- Don't allow children to poke or prod animals with sticks or other hard objects. Many of the animals that live in tide pools are soft and delicate and they may be injured.
- Instruct youngsters not to remove clinging animals such as sea anemones or sea stars (starfish) from rocks. These animals have soft body parts and they could be injured by trying to tear them off their rocks.
- Be careful where children walk. Advise children to look at where they place their feet and try not to step on any creatures or plants clinging to the rocks and ledges.
- Be aware of any regulations or laws protecting tide pool life. For example, many tide pools in California and Oregon are located in parks and preserves that are protected by state laws.

Here are three of the many organisms children can discover at the seashore (see related literature for descriptions of other seashore organisms):

Barnacles
What they look like:
Barnacles have shells of connected overlapping plates. They are $\frac{1}{2}$-inch to 1-inch wide and are volcano-shaped. Depending on the species, they may be white, brown, pink, or black. When under water they reach out with feathery barbed legs to strain out plankton from the water and to absorb oxygen.

Where they live:
They can be found all along the Pacific shoreline from Alaska southward to Mexico. They also inhabit the Atlantic seacoast from Canada south to Florida.

How to find them:
Barnacles glue themselves to rocks, ships, pilings, and even living creatures such as whales and abalones. Their habitats vary from exposed shorelines to protected bays. Typically they can be found in large and expansive colonies with hundreds or thousands of them in one place.

Sea Stars (Starfish)
What they look like:
Sea stars come in a wide variety of colors ranging from browns, blacks, reds, yellows, and oranges. Their outer skin is usually rough in appearance. Some species may appear to have short, knobby spines all over their outer surface. Most sea stars have five legs; a few varieties have as many as fifty legs. On the underside of a sea star are rows of tiny tubes (called tube feet), which are actually hollow muscular cylinders filled with water.

The Ochre Star (also known as the Common Sea Star) grows to 12 inches across. It is purple or orange with rows of white-tipped spines over its body. The Bat Star is webbed between its arms. It comes in a variety of colors—red, orange, brown, yellow, green, or purple. The Knobby Sea Star is brown with white spines. It grows up to 16 inches across.

Where they live:
There are more than five thousand different species of sea stars. They can be found in tide pools from Alaska to Baja California; and the Arctic to the Gulf of Mexico.

How to find them:
Sea stars cling under lower rock ledges, on rocky shelves, on seaweed mats, and in tiny crevices throughout a tide pool. Whenever there is a collection of mussels or other bivalves (two-shelled animals) sea stars are sure to be around.

Kelp
What it looks like:
Kelp looks like a large weed growing in the water. However, they are not flowering plants, but brown algae. Instead of roots they have a holdfast that anchors them to sand or rocks. What looks like a stem is called a stipe. The blades are similar to the leaves of a land plant. Some seaweeds have little gas-filled bulbs, called bladders, that help them float.

Where it lives:
Kelp (often referred to as California Giant Kelp) is found primarily along the coast of California. Other types of kelp, such as Horsetail Kelp, Sugar Kelp, and Edible Kelp, grow along the Atlantic seaboard from Maine to Rhode Island.

How to find it:
Kelp can be located along most rocky shores where it grows attached to rocks, rocky ledges, shells, and other hard surfaces. Often, after a storm, long strands of kelp can be found washed up on a beach.

Special Projects

A. Collecting and drying seaweed can be a great way for children to learn about life in and around tide pools. The many varieties of seaweed can be dried, flattened, labeled, and assembled into an interesting collection and re-membrance of kids' trip to a tide pool. Youngsters may wish to consult a good seaweed book to learn more about the varieties collected along the seashore. (*Note:* This project requires adult guidance and supervision.)

What Is Needed:
- hand or machine drill
- two sheets of plywood or fiberboard, 10 x 13 inches
- heavy white paper such as oaktag or drawing paper
- pieces of an old bed sheet
- lots of newspapers
- four long bolts with wing nuts
- two 8$\frac{1}{2}$ x 11–inch pieces of cardboard (cut from a box)

What to Do:
1. Drill holes (to fit your bolts) in each corner of the two sections of wood. The holes should be about 1 inch in from the top and side of each corner (drill the two boards together and the holes will be sure to match).
2. Put the connecting bolts through one board and lay it down so that the bolts stick upward.
3. Lay a piece of cardboard on top of this board.
4. Wet a sheet of heavy white paper and lay it on top of the cardboard.
5. Place a seaweed sample on top of the white paper.
6. Cover the seaweed with a piece of cloth.
7. Cover the cloth with a thick layer of newspaper.
8. Repeat steps 4 through 7 with other seaweed samples until children have mounted all the specimens they want to press.
9. Lay a second piece of cardboard on top of this pile.
10. Put the second piece of plywood on top of the last cardboard piece on the stack, threading the bolts through the holes.
11. Put a wing nut on each bolt and tighten them until pressure is felt. Then carefully tighten each bolt in turn as much as possible, putting even pressure on the stack.
12. Replace the wet newspapers with dry ones at least once each day.
13. After a week, carefully remove all the specimens from the press (the seaweed will be stuck to the white paper by a natural adhesive in the seaweed).

14. Invite children to label each of the specimens with its name, where it was collected, and the date of collection.

B. Examining the wide variety of plants and animals that live in a tide pool can be made easier with the use of a homemade tide pool viewer. This simple devise allows children to watch the activity in a tide pool without disturbing the occupants. (*Note:* This project requires adult guidance and supervision.)

What Is Needed:
- a no. 10 can (a three-pound coffee can is ideal)
- clear plastic food wrap (e.g., Saran Wrap)
- black flat paint and paint brush
- plastic or cloth tape
- one or more heavy rubber bands

What to Do:
1. Cut both ends off the can. Remove the contents and place them in another container.
2. Paint the inside of the can with black paint and allow it to dry overnight.
3. Use the plastic tape to wrap around the edge on one end of the can. Be sure to cover all the sharp edges.
4. Place the plastic food wrap over the other end of the can. Use one or more heavy rubber bands to keep the plastic wrap on the can (pull the plastic food wrap tightly so that it does not have any wrinkles). (*Note:* It usually takes two people to do this step.)
5. When children visit a tide pool, have them carefully place the end of the can that has the plastic food wrap on it a few inches into the tide pool. View the occupants through the other end of the can.

Incredible Facts

The "glue" that holds a barnacle to rocks is one of the strongest known natural adhesives.
- *Why would a barnacle need such a strong adhesive?*
- *What other animals need to stick to a surface?*
- *What animals "stick" to other animals?*

A sea star (starfish) can travel at a speed of about 2 to 3 inches a minute.
- *How does a sea star use its arms to travel?*
- *How far can a starfish travel in an hour?*
- *How fast can you travel in an hour?*

A chemical in kelp, known as algin, is used in chocolate milk, salad dressings, puddings, and ice cream.
- *What other chemicals do we get naturally from nature?*
- *What are the other ingredients in chocolate milk, pudding, and ice cream?*
- *How much ice cream is sold in the United States in one year?*

Sand dollars eat grains of sand to stabilize themselves in moving water.
- *Where can you find sand dollars?*
- *How did sand dollars get their name?*
- *What is the world's largest sand dollar?*

Related Literature

Burnie, David. *Seashore*. (New York: Dorling Kindersley, 1994).
A richly photographed and detailed examination of life at the seashore.

Chinery, Michael. *Questions and Answers about Seashore Animals*. (New York: Kingfisher, 1994).
More than two dozen questions and answers about some common and not-so-common seashore animals.

Lazier, Christine. *Seashore Life*. (Ossining, NY: Young Discovery Library, 1991).
An accurate overview of seashore life presented in a direct and factual format.

Rinard, Judith. *Along a Rocky Shore*. (Washington, D.C.: National Geographic Society, 1990).
Illustrated with crisp photographs, this book introduces the reader to life along the rocky sea coast.

Following are additional books in the *One Small Square* series by Donald M. Silver:

African Savanna
Arctic Tundra
Backyard
Cactus Desert
Cave
Pond

Chapter 19

Peter's Place

Sally Grindley

SAN DIEGO: GULLIVER, 1996

Book Summary

Every day, Peter plays on the shores beneath his hillside home. Animals of every type inhabit the waters below and birds of every description flock to Peter for bits of food. But one night a tanker crashes into the jagged rocks offshore, spilling foul-smelling oil into the water and across the beach. Peter washes, cleans, and scrubs both the land and his beloved animals, hoping he can return them back to some sense of normalcy.

Questions to Share

1. How are Peter's thoughts about the environment similar to or different from your thoughts about the environment in which you live?
2. How do you think the illustrations complement the story?
3. How are the events in this book similar to any events that have happened in real life?
4. What do you think was the greatest threat to the animals in this story?
5. How can events in this story be prevented in real life?

Major Book Project

Oil pollution from grounded tankers is an increasing problem throughout the world. Although this story takes place in a remote spot in England, the repercussions of the oil spill can spread up and down the coast and last for many years. Hardly a year goes by that there is not some major oil spill somewhere in the world. The following activity alerts children to the speed at which a seaside community can become fouled by oil.

Provide children with four sealable sandwich bags. Label the bags "A," "B," "C," and "D." Fill each bag $1/3$ full with water and $1/3$ full with used motor oil. Invite children to place a hard-boiled egg in each bag. Seal the bags. Have youngsters remove the eggs from each of the bags (they should wear kitchen gloves or some sort of disposable gloves) according to the following schedule:

From Bag "A"—After 15 minutes
From Bag "B"—After 30 minutes

From Bag "C"—After 60 minutes
From Bag "D"—After 120 minutes

Encourage children to peel each of the hard-boiled eggs and note the amount of pollution that has seeped through the shell and into the actual egg. Which egg has the most pollution? How rapidly did the pollution seep into each egg? Set aside some time afterward to discuss the rapidity with which these eggs became polluted and the implications for spilled oil polluting a beach or shoreline.

For an extension of this activity, invite youngsters to use a variety of soaps and detergents to wash off the oil from these polluted eggs. Which soap/detergent works best? What difficulties do they encounter when trying to clean the eggs? How long does it take to clean an egg?

As a further extension, invite children to dip an old, discarded stuffed animal in a pail of water in which some motor oil has been added. Encourage them to attempt to clean off the oil with a variety of soaps and detergents (have them wear gloves). Again, what difficulties do they have and how long does the cleaning process take?

Write Away

Several organizations have brochures, leaflets, and guidebooks on ocean pollution and ways to prevent it. Encourage youngsters to write to several of the following organizations and request pertinent information. When the resources arrive, plan time to discuss the methods and procedures in which the children can participate to prevent or alleviate this global problem. Invite them to prepare an "action plan" for themselves and their friends in which they take a proactive stance against ocean pollution.

The New York Sea Grant Extension Program
SUNY
125 Nassau Hall
Stony Brook, NY 11794-5002
(516) 632-8730
This program has a 24-page booklet entitled "Earth Guide: 88 Action Tips for Cleaner Water"; copies are free.

NOAA Marine Debris Information Office
Center for Marine Conservation
1725 DeSales Street NW
Washington, D.C. 20036
A variety of informational brochures are available from this group.

International Wildlife Coalition
Whale Adoption Project
634 North Falmouth Highway, Box 388
North Falmouth, MA 02566

Have youngsters write to this organization if they are interested in "adopting" an endangered animal—specifically a whale.

Strathmore Legacy's Eco Amigos Club
333 Park Street
West Springfield, MA 01089
Invite children to write to this coalition of environmentally friendly youngsters from around the country.

Keep America Beautiful
99 Park Avenue
New York, NY 10016
Have children ask for "Pollution Pointers for Elementary Students," a list of environment improvement activities.

Across the Curriculum

Geography

Invite youngsters to watch the local news or read the local newspaper for reports of ocean pollution from around the world. While they may wish to focus on events related to grounded tankers, other types of pollution can be tracked as well. Invite children to hang up a large wall map of the world. For each incidence of ocean pollution, invite youngsters to write a brief summary (date, nature of occurrence, place, resolution, etc.) on a 3 x 5 index card. Post each card around the wall map and connect the card with the actual location on the map using a length of yarn (the yarn can be taped or pinned to the wall). Encourage youngsters to maintain this map over a long period of time (6 months to a year) and note the number of incidents, where most of them occur, the severity of each incident, the amount of shoreline polluted and the estimated number of wildlife killed or threatened. Plan time to discuss the short-term and long-term implications of these events.

Math

As youngsters record the events above, invite them to keep a running tally of the total number of gallons of oil discharged into the ocean. Encourage children to create a line graph separated into the months of the year. Invite them to plot the number of gallons discharged each month on the graph and connect these points with a line. In which month was the largest amount of oil discharged into the ocean? Would these totals be expected in succeeding years?

Experiment

The book discussed some of the efforts to contain the spreading oil. One of the illustrations showed a boom (a floating line) being placed in the water in an attempt to contain the oil spill. Children can experiment with different devices to contain their own oil spill to determine the best material to use in a boom.

Place water in a large round pan or pie plate. Invite youngsters to collect several different floating objects that could be used as booms. (These may include a rubber band; a length of yarn, string, twine, or cotton batting; a ring of styrofoam cut from the top of a disposable coffee cup, etc. Invite children to use their creative powers in inventing other types of potential booms.) Place several drops of cooking oil in the middle of the water (children will note that since oil and water do not mix, the oil floats on top of the water). Invite children to encircle the oil with one or more different devices to determine which device or which material best prevents the oil from spreading across the surface of the water.

To further test their devices create small ripples in the water of the pan with your finger. Invite children to note what happens to the oil on the surface. Take time to discuss the difficulties that arise in the ocean when the surf is high or the seas are rough and there is a need to contain oil on the surface. How do their devices work in containing oil on a choppy "sea"? What are the implications in real-life rescue efforts?

Incredible Facts

Each year in the United States, 2.1 million tons of used motor oil is spilled into rivers and streams.
- *How many gallons are in one ton of oil?*
- *How many gallons are spilled in one decade? How many are spilled in a century?*

A single quart of motor oil can pollute 250,000 gallons of drinking water.
- *How many gallons of water does your family use every day?*
- *How much of that water is used for drinking purposes?*
- *How much water does your family use in a year?*

Each year about 2.3 trillion gallons of liquid wastes are discharged directly into U.S. coastal waters.
- *Which states discharge the most liquid wastes?*
- *Which states are most in danger from discharged liquid wastes?*
- *What are the effects on wildlife from discharged liquid wastes?*

One pint of motor oil can create an oil slick an acre in diameter.
- *How large is an acre?*
- *How large an oil slick could be created with one gallon of oil? How large with one hundred gallons?*

Related Literature

Hare, Tony. *Polluting the Sea.* (New York: Gloucester Press, 1991).
A useful source for projects, this information-packed book examines the damage caused by oil spills and other pollutants.

Hoff, Mary. *Our Endangered Planet: Rivers and Lakes.* (Minneapolis, MN: Lerner, 1991).
This book clearly points out the dangers of pollutants to fresh bodies of water and actions that can prevent these problems.

Javna, John. *50 Simple Things Kids Can Do to Save the Earth.* (Kansas City: Andrews and McMeel, 1990).
A compendium of Earth-friendly ideas that encourage kids to get involved in their home planet.

Spurgeon, Richard. *Ecology.* (Saffron Hill, England: Usborne Publishing, 1988).
A magnificent introduction to ecology, the inter-relationships between humans and their environment, and ways youngsters can become involved in its preservation.

Chapter 20

A Swim Through the Sea

Kristen Joy Pratt

NEVADA CITY, CA: DAWN PUBLICATIONS, 1994

Book Summary

Seamore the Seahorse explores the fascinating plants and animals of the undersea world. His alphabetic journey takes him past a flashlight fish, some manatees, a porcupine fish, and a wise and wondrous whale, among others. This journey through the undersea world is a beautiful environmental awareness book full of interesting facts and colorful illustrations. Equally important is the fact that it was written and illustrated by a 17-year-old high school student.

Questions to Share

1. How did this book help you become aware of the wide variety of life in the ocean?
2. Which of the creatures did you enjoy the most? Which did you find most amazing or unusual?
3. Based upon what you learned in this book, would you like to learn more about life in the ocean?
4. If you had an opportunity to talk with the author/illustrator, what would you like to discuss?

Major Book Project

The oceans of the world are rich with a diversity of life. Many different people around the world depend on the oceans for their livelihood. Provide children with a list of oceanic occupations similar to the ones listed on the next page. Invite them to conduct some library research on one or more of the chosen occupations with specific reference to job requirements, training and education, amount of time spent at sea, average salary, long-term prospects for the occupation, and occupational dangers. Provide opportunities for children to share their research with others (if you are working with a group or class of children, they may wish to gather their information together into an informative brochure or booklet for display in the school or local public library).

Commercial fishers catch ocean creatures to sell to markets.
Marine geologists study rocks and the formation of the ocean floor.
Marine biologists study animal and plant life in the ocean.
Divers assist in finding sunken treasures, repairing underwater equipment, gathering information for research, and so forth.
Oceanographers study and explore the ocean.
Offshore drillers explore beneath the ocean floor for deposits of petroleum and natural gas to be used for various forms of energy.
Mariculturists raise farm fish and other sea life for food and/or restocking the ocean.
Marine ichthyologists study fish, their habitats, the food they eat, and their relationship to their environment.
Marine ecologists study the relationships between sea creatures and their environment (e.g., the effects of pollution on a particular species).
Captain/crew of ship works on a commercial boat or cruise ship.
Navigators use directions to determine a ship's course at sea.

Encourage youngsters to discover more about one or more of these occupations and report their findings to others or prepare a short report entitled, "A Week in the Life of a _____." Encourage youngsters to give some thought to how their chosen ocean occupation might change in the future.

Write Away

Invite children to write to one or more of the following organizations for information and data related to ocean life. Encourage them to pay particular attention to any data related to the preservation of the world's ocean—with particular emphasis on pollution control and/or elimination.

Cetacean Society International
P.O. Box 9145
Wethersfield, CT 06109

Cousteau Society
Greenbriar Tower II
870 Greenbriar Circle, Suite 402
Chesapeake, VA 23320

International Whaling Commission
The Red House
135 Station Road
Histon, Cambridge
England CB4 4NP

What's Cooking?

Children are often amazed at the incredible variety of food harvested from the world's oceans. You and the children with whom you work may wish to put together an "ocean picnic" using ocean-related recipes (see below) or items that simulate ocean life. This feast can be shared with family members or as a classroom activity. Plan to discuss the various types of food harvested from the ocean and provide children with library resources for learning more about those food items.

Tasty Crab Dip

Ingredients

7 oz. can crabmeat
8 oz. package cream cheese
1/4 cup cocktail sauce

Directions

Mix together in a small bowl. Serve with crackers and enjoy.

Ocean in a Cup

Ingredients

2 large packages blueberry jello
1 package small gummy fish
Graham cracker crumbs
beverage umbrellas

Directions

Mix jello according to package directions. Pour into clear plastic cups leaving a 1-inch space at top. Refrigerate until slightly set. Mix gummy fish into each cup. Refrigerate until firm. Put graham cracker crumbs on half of each "ocean" to simulate beach. Insert beverage umbrella into "sand."

Shark Eggs

Ingredients

6 hard-boiled eggs
4 oz. can drained tuna
mayonnaise and pickle relish
paprika

Directions

Slice eggs in half lengthwise. Carefully scoop out yolks and put in mixing bowl. Add tuna, mayonnaise, and pickle relish to taste in a bowl. Stir until blended. Fill egg centers with mixture and sprinkle with paprika.

Sea Mix

Ingredients

1 package mini-fish pretzels
1 package mini fish crackers
   (cheese)
1 package fish cookies
1/2 lb. gummy fish
1 can nuts
1 box raisins

Directions

Combine all ingredients. Serve in a sand pail.

Across the Curriculum

Language Arts

Invite youngsters to create their own poetic view of one of the creatures in this book. Present them with a diamonte poem similar to the one below.

Octopus	(one word)
Speedy, shy	(two words)
Gliding, hunting, creeping	(three words)
Arms, beak, suckers, tongue	(four words)
Feeding, tasting, walking	(three words)
Mysterious, clever	(two words)
Invertebrate	(one word)

Provide youngsters with the following form for a diamonte, inviting them to select an animal from the book and create one or more poems for sharing.

Name of animal
Adjective, adjective
Participle, participle, participle
Noun, noun, noun, noun
Participle, participle, participle
Adjective, adjective
Classification of animal

Incredible Facts

⟿ **Bottlenose dolphins produce sounds with frequencies of about 200,000 vibrations per second—about the same as those produced by bats.**
   • *Why do animals such as dolphins and bats produce sound vibrations?*
   • *How does human hearing compare with that of other animals?*
   • *Is the ability to hear many sounds related to intelligence?*

When grouper fish spring into action, the first stroke of the tail is so powerful that it creates a sonic boom.
- *Why do groupers have such powerful tails?*
- *What other animals have powerful tails?*
- *What are some of the ways in which animals use their tails?*

 The tentacles of a pink jellyfish can be 100 feet long.
- *What would be the advantage of having 100-foot tentacles?*
- *What other animals have long arms?*
- *What animal has the longest arms in the animal kingdom?*

 The seahorse can beat its fins up to 30 times a second to propel itself through the water.
- *How fast does a seahorse travel through the water?*
- *What is the fastest sea animal?*
- *What is the fastest land animal?*
- *What is the fastest animal in the animal kingdom?*

Related Literature

Hopkins, Lee Bennett. *The Sea Is Calling Me.* (San Diego, CA: Harcourt Brace, 1986).
A collection of twenty-one poems about the sea.

Lauber, Patricia. *Who Eats What?: Food Chain and Food Webs.* (New York: HarperCollins, 1995).
A beautifully illustrated book that explains the concept of a food chain and how plants, animals, and humans are ecologically linked.

Lauber, Patricia. *An Octopus Is Amazing.* (New York: HarperCollins, 1990).
An introduction to one of the curiosities of the sea—the multitentacled, highly intelligent octopus.

Pallota, Jerry. *The Ocean Alphabet Book.* (Watertown, MA: Charlesbridge, 1989).
An introduction to various ocean creatures—one for each letter of the alphabet.

Wu, Herbert. *Life in the Oceans.* (Boston: Little, Brown, 1991).
Stunning photographs highlight this examination of the wide variety of ocean life throughout the world.

Zim, Herbert. *Seashore.* (New York: Golden Books, 1990).
This book investigates the plants and animal life found at the seashore and suggests related experiments and activities.

Part VII

Weather—
Weather or Not

Chapter 21

Bringing the Rain to Kapiti Plain

Verna Aardema

NEW YORK: DIAL BOOKS, 1981

Book Summary

A wonderful African folktale with a rhythmic pattern and an engaging story line. This story involves Ki-pat, a herdsman who watches his cows go hungry and thirsty because there is no rain to make the grass grow. He watches a large cloud overhead and devises an ingenious way to make the rain come using a special arrow and a special feather. A delightful story well told and equally well illustrated.

Questions to Share

1. How did the illustrations help you appreciate this story?
2. If you had an opportunity to talk with Ki-pat, what would you like to say?
3. Are the conditions in this part of Africa similar to any conditions in the United States?
4. Does the rhyming pattern of this story remind you of other rhymes?
5. What did you enjoy most about this book?

Major Book Project

Most children have seen rain fall. Most have heard the patter of raindrops on a roof. But, few children have had an opportunity to collect and examine raindrops up close. The following activity provides youngsters with a chance to see what a raindrop looks like.

Invite children to fill a cake pan (9 x 13) with about 2 inches of sifted all-purpose flour. At the start of a rainstorm, encourage kids to stand outside for a few moments and collect approximately thirty to fifty raindrops in the flour. (This will have to be done quickly and at the beginning of the rainstorm. A few moments in the rain will be sufficient.) Have them bring their pans inside and allow them to sit overnight.

The next day, encourage children to gently sift through the flour and gather up all the congealed raindrops they can. Have them organize the drops on a flat surface (the kitchen counter). They will note that raindrops come in two basic sizes—large and small (large raindrops fall from higher clouds and thus gather more moisture than small raindrops). Children will also note that the drops will be in a state of expansion or contraction. (As a drop falls it expands and contracts—a natural atomic reaction. Depending on when it hits the pan of flour, it will either be in a state of expansion or a state of contraction.)

Invite children to organize their drops by size (large vs. small) or by state of expansion or contraction. Invite them to develop ratios on the number of large drops and the number of small drops. If desired, charts and graphs of this information can be constructed. Based on the size of the majority of the drops, children can make inferences about the distance the drops fell during the beginning of the storm.

Children may wish to preserve their raindrops. This can be done by spraying or carefully dipping the drops with a commercial varnish. When the raindrops are dry, youngsters will be able to manipulate the drops in a variety of activities.

Creative Dramatics

A. Read the story aloud to children. Afterward, invite them to read along with you. Read the book again and pause at the beginning of each of the repeated phrases. Invite children to "fill in the blanks" with words remembered from the story.

B. After reading the story to children, read it again. This time encourage children to clap their hands or pat their legs for each syllable in the repeated phrases.

C. If you are working with a group of children, assign a repeated line to each of several different children. Read the story aloud again and invite kids (one at a time) to chant their identified line in cumulative fashion.

Across the Curriculum

Language Arts

After reading the story to children invite them to create their own "folktale" about the setting or events in their town. Encourage them to use the rhythmic pattern in the book. Following are three examples that children can modify according to their own locale:

> This is the great Colorado plain,
> All dry and coarse without the rain—
> With rocks and boulders scattered around,
> And distant birds with a crying sound.

> These are the great Pennsylvania hills
> Sprinkled with farms and streams and mills—
> An ocean of trees for birds to soar through,
> And waves of flowers for insects to fly to.

This is the great Oregon city,
Asleep by the ocean, quiet and pretty—
Encircled by sand and washed by tides,
And brushed by wind on all its sides.

Encourage children to continue with their own cumulative tales about their specific part of the world. Youngsters may also elect to write a similar cumulative tale about a place they have visited on vacation or a favorite location they would like to go to. Provide opportunities for children to share their tales.

Experiments

A. Invite students to place several ice cubes on a cookie sheet. Using kitchen mitts hold the sheet over a pot of boiling water or the spout of a tea kettle (*Note:* This should be done by an adult, not children). Water will begin to form on the bottom of the cookie sheet. Invite children to observe and offer possible explanations for what they see. How is this demonstration similar to how rainclouds form?

B. Invite youngsters to watch as you fill an empty mayonnaise jar with boiling water. Allow the jar to sit for several minutes and then pour off about half of the water. Place the lid of the jar upside down on top of the jar. Put several ice cubes on the lid. Invite children to note the drops of water forming on the underside of the lid. Again, invite them to speculate on the similarity of this activity and the formation of rain in the atmosphere. (Warm air holds more moisture than cold air. When moisture-laden air is cooled [as in these two demonstrations] it loses that moisture and "rains.")

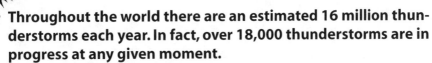

Throughout the world there are an estimated 16 million thunderstorms each year. In fact, over 18,000 thunderstorms are in progress at any given moment.
- *How many thunderstorms occur in your area each year?*
- *How much rain falls from an average thunderstorm?*
- *What part of the world has the most thunderstorms each year?*

A typical thunderstorm has the energy of ten atomic bombs. How much energy is in one atomic bomb?
- *How much lightning occurs in an average thunderstorm?*

During a typical rainstorm about 5 million drops of water fall on an acre.
- *How many drops of water are in a gallon?*
- *How many gallons of water fall on an acre?*
- *How many acres receive rain during a typical rainstorm?*

Little raindrops fall at a speed of 6 miles per hour; large raindrops fall at a speed of 18 miles per hour.
- *How fast can you run 100 yards (in miles per hour)?*
- *Can you run as fast as a large raindrop?*
- *How fast does hail, snow, or sleet fall?*

Related Literature

Markle, Sandra. *A Rainy Day.* (New York: Orchard, 1993).
An illustrated explanation of why it rains, what happens when it does, and where the rain goes.

Martin, Bill, and John Archambault. *Listen to the Rain.* (New York: Henry Holt, 1988).
A playful and creative introduction to the sounds of rain.

Otto, Carol. *That Sky, that Rain.* (New York: Crowell, 1990).
An introduction to the water cycle with an emphasis on the creation of rain.

Polacco, Patricia. *Thunder Cake.* (New York: Philomel, 1990).
An approaching storm frightens a young girl until her grandmother shows her how to bake a thunder cake.

Serfozo, Mary. *Rain Talk.* (New York: McElderry Books, 1990).
During a summer rainstorm, a small girl notices the way the rain affects her senses.

Steele, Philip. *Rain: Causes and Effects.* (New York: Watts, 1991).
An excellent introduction to rain and its effects on humans. Includes several simple experiments.

Chapter 22

Cloudy with a Chance of Meatballs

Judi Barrett

New York: Macmillan, 1978

Book Summary

Across an ocean, over some bumpy mountains and three hot deserts lies the tiny town of Chewandswallow. Although this town looks like any other small town—it isn't. For in the town of Chew-andswallow it rains food three times a day. Hot dogs, hamburgers, mashed potatoes, eggs, and toast sprinkle the town and all the people have to do is go outside and collect their meals. But, then, something happens and the town is never the same again. A marvelously inventive story that offers readers a humorous and creative way of looking at weather.

Questions to Share

1. If it could rain food where you live, what food would you most like to have?
2. Other than where you live now, where would you like to live in the world?
3. What was the most unusual type of weather in the town of Chewandswallow?
4. Why do you think the illustrator made the drawings at the beginning and end of the story in black and white and those in the middle of the story in color?
5. What made this such a humorous story?

Major Book Project

Invite children to create their own homemade weather station. The following instruments will help them learn about the weather and some of the ways in which meteorologists measure various aspects of the weather.

Thermometer

Use a nail to dig out a hole in the center of a small cork. Fill a bottle to the brim with colored water and push the cork into the neck of the bottle. Push a straw into the hole in the center of the cork. Mark the line the water rises to in the straw with a felt-tip pen. Note the temperature on a regular thermometer and mark that on a narrow strip of paper glued next to the straw. Take measurements over several days, noting the temperature on a regular thermometer and marking that at the spot where the water rises in the straw on the strip of paper. After several readings, youngsters will have a fairly accurate thermometer. (Liquids expand when heated [water rises in the straw] and contract when cooled [water lowers in the straw].)

Barometer

Stretch a balloon over the top of a widemouthed jar and secure it with a rubber band. Glue a straw horizontally on top of the stretched balloon starting from the center of the balloon (the straw should extend beyond the edge of the jar). Attach a pin toward the end of the straw. Place another straw in a spool and attach an index card to the end. Place this device next to the jar so that the pin is close to and points to the index card. (When air pressure increases, the pressure inside the bottle is less than that of the outside air. Therefore, the balloon rubber pushes down, and the pointer end of the straw moves up [that spot can be marked on the index card.]) When the air pressure goes down, the air inside the jar presses harder than the outside air. The rubber pushes up and tightens, and the pointer moves down. (Point out to children that when the pointer moves down, bad weather is probably on the way because air pressure falls when a storm is approaching. When the pointer rises, that's usually a sign that good weather is on the way.)

Anemometer

Cut out two strips of cardboard approximately 2 x 16 inches. Make a slit in the middle of each one so that they fit together to make an "X." Cut four small paper cups so that they are all about 1 inch high. Staple the bottom of each cup to one "arm" of the "X." Use a felt-tip marker to color one of the cups. Make a hole in the center of the "X" with a needle.

Stick the eye of a needle into the eraser of a pencil and place the pencil into a spool (jam some paper around the sides of the spool hole so that the pencil stays erect). Glue the spool to a large block of wood. Place the "X" on the tip of the needle so that it twirls around freely. Blow on the cups to make sure they spin around freely (some adjustments on the size of the hole may need to be made). Invite youngsters to place their anemometer outside on a breezy day and count the number of times the colored cup spins past a certain point. That will give them a rough idea of wind speed. (Meteorologists use a device similar to this one, but the revolutions are counted electronically.) Later, you may wish to introduce youngsters to the Beaufort Scale—a widely used scale to judge the speed of wind.

Children may wish to set up their makeshift weather station outside and take regular "readings." These measurements can be matched with those reported in the daily newspaper. Comparisons between the readings obtained by children and those reported in the daily paper can be discussed. Children should record their readings over a period of several days or weeks in a journal or appropriate notebook.

Special Projects

A. Throughout the ages people have been fascinated by the weather. As a result, many sayings and predictions have been handed down from one generation to the next. These sayings have helped us to understand more about the weather and to appreciate some of the difficulties we may have in controlling it. Here are two sayings or admonitions that have been passed down through the years:

> "Red sky at morning—sailor take warning.
> Red sky at night—sailor's delight."

> "A January fog will freeze a hog."

Invite youngsters to look through other books and assemble a collection of weather sayings that have been handed down through the years. How accurate are those sayings? How do those sayings compare with actual meteorological events?

B. Because people did not always understand the weather, they have had many misconceptions or strange beliefs about the conditions or situations that caused many weather patterns. Following are a few that people have had:

- Sea fog was once thought to be the breath of an underwater monster.
- The ancient Chinese thought that storms were caused by dragons fighting in the sky.
- The Aztecs believed that the Sun god could only be kept strong and bright through the use of human sacrifices.
- The Norse thought that weather was created by the god Thor, who raced across the sky in a chariot pulled by two giant goats.

Invite youngsters to research other beliefs that people had about the weather. They may wish to collect their data from trade books, encyclopedias, or conversations with weather experts. Encourage them to put together a collection of these beliefs into a notebook or journal.

C. Many strange things have fallen from the sky as a result of unusual weather patterns. Kids may want to research some of these unusual events and assemble them together into a notebook of "Weird and Wacky Weather." Here are a few to get them started:

- On October 14, 1755, red snow fell on the Alps.
- In June 1940 a shower of silver coins fell on the town of Gorky, Russia.
- On June 16, 1939, it rained frogs at Trowbridge, England.

What's Cooking?

Invite youngsters to reread the book and "shop" from the sky. They may wish to prepare shopping lists of food words selected from the story and arrange them into food groups or according to how those products are located in the local supermarket. Food words can be written on separate index cards and then placed inside brown lunch bags according to individual categories.

Invite youngsters to assemble the food words into various menus for family members. Are they able to put together a series of balanced meals for several days or an entire week? Are they able to provide for all the dietary needs of family members?

Across the Curriculum

Language Arts

Invite children to create sayings and phrases similar to those used in the book to describe the weather conditions in their own locality over a period of several weeks or months. Encourage youngsters to make their sayings as wild and imaginative as those described in the story (each weather condition is matched with a specific type of food). Following may be some examples: "The sun was like a melted lump of butter"; "Rain came down like spraying soda pop"; and "The snow was like huge mountains of mashed potatoes." Discuss with youngsters the various ways we describe the weather and its similarities with different foods. Invite youngsters to create an ongoing encyclopedia of weather sayings and phrases.

Incredible Facts

Snow has so much air in it that 10 inches of snow will melt down to 1 inch of water.
- *What part of the United States gets the most snow each year?*
- *What is the largest snowfall ever recorded?*
- *If all that snow was converted to water, how much water would there be?*

In one year the city of Cherrapunji, India, received 1,042 inches of rain.
- *How many yards of rain is that?*
- *How much more rain did Cherrapunji receive in that year than your city did last year?*
- *Why does that part of India get so much rain?*

On September 3, 1970, a hailstone fell on Coffeyville, Kansas, that was 17 inches round and weighed 1.7 pounds.
- *What is the size of an average hailstone?*
- *How much larger was the Coffeyville hailstone than an average hailstone?*
- *What causes hailstones?*

On April 12, 1934, a gust of wind blowing at 231 miles per hour was recorded on Mount Washington in New Hampshire.
- *How fast do the winds in a hurricane blow?*
- *Why is Mount Washington so windy?*
- *What is the windiest city in the United States?*

Related Literature

Gibbons, Gail. *Weather Forecasting*. (New York: Four Winds, 1987).
A simple, yet thorough introduction to the importance and significance of weather forecasting.

Leslie, Clare. *Nature All Year Long*. (New York: Greenwillow, 1991).
A month-by-month seasonal guide about changes in plant and animal behavior.

McMillan, Bruce. *The Weather Sky*. (New York: Farrar, Straus and Giroux, 1991).
This book presents a year's worth of sky changes along with color photographs and descriptive illustrations.

McVey, Vicki. *The Sierra Club Book of Weatherwisdom*. (San Francisco, CA: Sierra Club, 1991).
Basic weather principles and experiments in tandem with weather customs and traditions highlight this book.

Wyatt, Valerie. *Weatherwatch*. (Reading, MA: Addison-Wesley, 1990).
A book packed with weather lore, weather facts, weather data, and loads of experiments.

Chapter 23

The Loon Spirit

Phil Harper
Minocqua, WI:
NorthWord Press, 1995

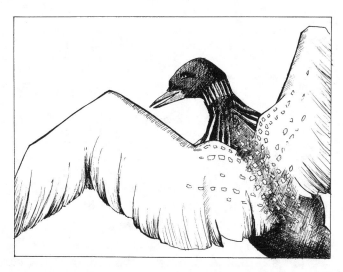

Book Summary

At the end of winter's reign over the frozen northwoods, Lakucha, the Loon Spirit, awakens. With a haunting cry that heralds the coming of spring, Lakucha's mighty wings unfurl, clearing the snow away from the hibernating wilderness. Rising above the northern world, the Loon Spirit soars over the land and brings warmth and new life. As nature awakens, Lakucha transforms into an ordinary loon, choosing a familiar resting place until fall. This magnificent book brings the changing of the seasons to life and underscores the inter-relationships that exist throughout nature.

(*Note:* This title is included in the "Weather" part of this book because of its emphasis on the changing seasons. Readers, however, will see many connections with "Environment," "Plants," "Animals," and "Earth." The beauty of this book is that it can be used throughout most science disciplines as well as across the entire elementary curriculum.)

Questions to Share

1. Why do people invent legends like this to tell about natural occurrences (like the changing seasons)?
2. How much of this story is fiction and how much is fact? Can you tell the difference between the two?
3. Why is the "coming of spring" such an important event?
4. What are some of the signs and symbols showing that spring has returned to the land?
5. Why does Lakucha change into a common loon?
6. Why does the author begin the story in winter and end the story in winter?

Major Book Project

The following activity will help youngsters understand why there are four seasons in most parts of the world.

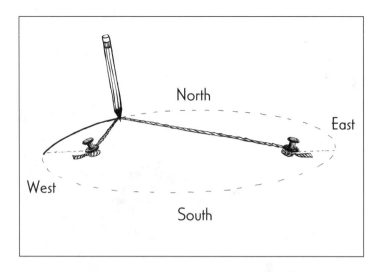

Push a knitting needle through a large rubber ball (the ball will represent the earth and the needle will represent the earth's imaginary axis). Draw an ellipse about 12 inches in diameter on a large piece of newsprint or paper (the ellipse represents the Earth's oval orbit).

To draw an ellipse, place a sheet of paper on a piece of cardboard. Push 2 pins about 2 inches apart into the paper (and cardboard). Tie 2 ends of a 6-inch length of string together and loop the string over the two pins (see the illustration). Insert a pencil in the loop and pull the string tight. Keeping the string tight, move the pencil in a circular motion and you will draw an ellipse.

Mark the four quarter points "North," "South," "East," and "West." Place an unshaded lamp in the center of the cardboard (this represents the sun).

Move the ball in turn to each of the four positions holding the needle straight up and down. Invite youngsters to observe which part of the ball is lit up.

Now tilt the ball so that the axis is slanted about 23.50° away from the vertical. Place the ball in turn at each of the four positions, keeping the needle slanted in the same direction. Invite youngsters to look at the lighted section of the ball. In each position, have them observe which part receives the direct rays and which receive the slanting rays of light.

(The axis of the Earth points to the North Star at a 23.50° slant. It is this slant that makes the seasons change as the Earth revolves around the sun. When the side we live on is tilted toward the sun, we have summer because we receive the direct rays of the sun. Six months later—when our part of the Earth is tilted away from the sun—it is winter because we receive the sun's rays at a slant and therefore get less of the sun's heat. At the equator, the sun's rays are always direct; thus, there are no seasons.)

Invite children to create their own models similar to the one above. They may wish to put a small mark on the ball (this represents their position on the surface of the Earth) and note whether the "sun's" rays are direct or slanted as the ball moves around the "sun." Have youngsters write the names of the seasons around the outside of the ellipse.

Special Project

Many cultures and peoples have legends about the seasons, the stars, the weather, or other mysterious happenings in nature that may be beyond the realm of comprehension or understanding. Legends helped "explain" why some things happened or why certain events occurred on a regular basis. The story of the Loon Spirit is how one group of people sought to explain the creation and maintenance of the seasons. Some of that story may be based in fact; other parts may be steeped in a certain amount of creativity and originality.

Invite youngsters to select books from the school or public library in which legends are used to explain seasons, weather conditions, stars and constellations, or forces of nature. Native American legends (*Star Tales: North American Indian Stories about the Stars* by G. Mayo [New York: Walker, 1987]) and Asian tales (*A Song of Stars: An Asian Legend* by T. Birdeye [New York: Holiday, 1990]) offer excellent examples of legends, myths, and stories handed down through the generations to explain the unexplainable. Invite children to gather a collection of books about the weather or about the seasons written in a "legendary" format. What similarities do they notice? What differences? Why are legends created in the first place? Why do these legends seem to stay around for long periods of time?

After youngsters have had an opportunity to share various legends from different cultures, invite them to create their own legend about the seasons. They should include some factual data to give the story a sense of authenticity as well as some creative and interpretive explanations of events. Provide opportunities for children to share these stories with others.

Arts and Crafts

Invite youngsters to give some thought to the colors that are associated with each of the seasons. For example, what colors would they associate with the season of winter? White (the color of snow) would be a natural selection; but they should also consider white plants, white animals, white objects. What other colors, besides white, could be used to describe winter? What colors would be used to describe summer? Orange—the color of the blazing sun; green—the color of deep forest trails; yellow—the color of birds flitting in and out of trees. Invite children to make charts or graphs of the colors associated with each of the seasons and the items or objects that display those colors.

While youngsters may be tempted to focus on colors of plants and animals, help them focus on other areas too. For example, what colors can be associated with a summer or spring sky? What does autumn dirt look like? What color is winter air? Invite children to look throughout their environment at the various colors and how their perception of a season may be shaped by the colors within that season. This can be a wonderful opportunity for kids to be more observant about the colors in their environment. Even children who live in regions of the country that do not experience the traditional four seasons will begin to notice that each season does have its own distinctive pattern of colors.

Across the Curriculum

Language Arts

Invite children to put together a collection of phrases or sayings related to the seasons. Examples could include "March comes in like a lion and goes out like a lamb" and "April showers bring May flowers." What other sayings or admonitions can they discover through library research and/or talking with others? How did some of those sayings come to be? How much truth is in each of those sayings?

Encourage youngsters to create additional sayings for seasons of the year for which they cannot locate appropriate "truths." Examples may include "Deep winter snows create wet springtime flows." These sayings or observations can be collected into a journal or notebook for sharing with others.

☔ **The average annual temperature worldwide is approximately 58°F.**
- *What is the average annual temperature where you live?*
- *What is the hottest temperature ever recorded where you live?*
- *What is the coldest temperature ever recorded where you live?*

☔ **The Zuni Indians of the American Southwest believed that a halo around the sun meant that rain was to follow shortly.**
- *What causes a halo to form around the sun?*
- *What conditions must be present for rain to occur?*
- *How often does it rain where you live?*

☔ **The ancient Chinese believed that dragons formed clouds with their breath and brought rain. Storms occurred when the dragons fought with one another.**
- *What causes a thunderstorm?*
- *How many thunderstorms occur in your area each year?*
- *How many thunderstorms occur around the world in a single day?*

☔ **The ancient Greeks believed there were eight winds, and each wind had its own name.**
- *What are some different types of wind we have?*
- *What type of wind is the strongest?*
- *What is the fastest type of wind that has ever occurred in your area?*

Related Literature

Asimov, Isaac. *Why Do We Have Different Seasons?* (Milwaukee, WI: Gareth Stevens, 1991).
Color photographs and illustrations describe the Earth's changing position in relationship to the sun.

Berger, Melvin. *Seasons.* (New York: Doubleday, 1990).
Rich narrative combines with scientific fact and large-scale pictures to offer lots of information for readers.

Branley, Franklyn. *Sunshine Makes the Seasons.* **(New York: HarperCollins, 1985).**
This book offers readers reasons why the seasons occur and change as they do.

Esbensen, Barbara. *Great Northern Diver: The Loon.* **(Boston: Little, Brown, 1990).**
An informative and descriptive account of one of nature's most amazing birds.

Klein, Tom. *Loon Magic for Kids.* **(Minocqua, WI: NorthWord Press, 1989).**
Lots of fascinating information and loads of colorful photographs distinguish this book about an intriguing bird.

Lerner, Carol. *Seasons of the Tall Grass Prairie.* **(New York: Morrow, 1980).**
A description of plant life on the American prairie, season by season.

Simon, Seymour. *Autumn Across America.* **(New York: Hyperion, 1993).**
This informational book describes the signs of autumn seen around America.

Tudor, Tasha. *Seasons of Delight.* **(New York: Philomel, 1986).**
A three-dimensional book that describes a year on an old-fashioned farm.

Part VIII

Space—
The Final Frontier

Chapter 24

Seeing Earth from Space

Patricia Lauber

NEW YORK: ORCHARD BOOKS, 1990

Book Summary

This book is a grand and marvelous examination of the planet Earth from far out in space. Using photographs from satellites, shuttle missions, and Apollo moon flights the author presents an intriguing and awe-inspiring view of the "big blue marble." Readers will be amazed at how the Earth looks from thousands of miles away and the incredible information that can be learned from these vantage points. An "eye-popping" book.

Questions to Share

1. Which of the photographs did you find most amazing? Why?
2. Based on this book, which part of the world would you like to see more of from space?
3. Would you want to travel into space like an astronaut?
4. How do the photographs in this book differ from those you take?
5. How are photographs like the ones in this book helpful to scientists?

Major Book Project

This book provides readers with some striking photographs of the planet Earth from a perspective that few youngsters have seen. The following activity is designed to help them appreciate the enormous amount of detail these photos reveal and how they have helped scientists learn more about the planet Earth.

Provide youngsters with an inexpensive disposable camera. Invite them to select an object or item in their yard or neighborhood that is at least 100 yards in the distance (i.e., a tree, a building, a house). Invite them to take a photograph of that object. Encourage them to take a series of continuing photos of the same object—each time moving closer to the object in 10-yard increments as follows:

1st photo—100 yards distant
2nd photo—90 yards distant

3rd photo—80 yards distant
4th photo—70 yards distant
5th photo—60 yards distant, etc.
(the last photograph can be taken directly in front of the object)

When the photographs are developed, invite youngsters to lay them on a table in the correct sequence (from most distant to least distant). Encourage kids to note any differences between any two photographs (e.g., what details are seen in the photo taken 30 yards away that were not noticed in the photo taken 40 yards away?). Children may wish to work together to scan each of the photographs and arrive at mutual decisions about the differences noted between photos. Later, encourage kids to talk about the differences noted and how their powers of observation are changed by the distance between the object and the camera.

Encourage children to assemble their collections of photographs into a scrapbook or notebook. A caption can be supplied for each photo identifying the items seen and how one photo differs from any of the ones that precede it.

Creative Dramatics

Gather a group of youngsters together in a circle and tie the end of a ball of twine to your finger. Toss the ball to one of the youngsters, who in turn wraps the string around his/her thumb and tosses it to another individual. Have children do this until all are joined in the circle with the twine. Have youngsters imagine that they are the Earth. Each child becomes a different part of the Earth (use examples from *Seeing Earth from Space* such as Africa, California, Madagascar, Hawaii, Algeria, and the Mississippi River). Each child tells how his/her area is affected by humans (e.g., poaching of wildlife, deforestation of old-growth timber, oil pollution in the Atlantic, and overfishing in the North Pacific). As each individual speaks, he/she tugs on the twine so that all individuals feel the effects. Once all the youngsters have spoken, repeat the process, but have each youngster tell one thing that can be done to improve his/her part of the Earth. After the game, have individuals react to what has been said and to the significance of the twine that connected them all. Take time to talk about some of the photographs in *Seeing Earth from Space* and how, although they may encompass large areas of the Earth's surface, they are related and connected to other parts of the Earth. This game, and any follow-up discussions, should assist youngsters in understanding the enormity of the third planet from the sun as well as the connections that exist between any two places on the surface of that planet.

Write Away

Children may wish to obtain some space-related literature from a variety of government organizations and agencies. Encourage them to request one or more of the following:

National Aeronautics and Space Administration
Educational Publications
Washington, D.C. 20546
Request the Astronaut Fact Book, which includes biographical sketches
    on current and former astronauts. Also available are free pictures about
    various space missions (i.e., "Apollo 17's View of Earth").

John F. Kennedy Space Center
NASA BOC-155
Kennedy Space Center, FL 32899
Ask for an information packet about the Space Center's function and ac-
    complishments.

Space and Rocket Center
One Tranquillity Base
Huntsville, AL 35807
Kids interested in learning about space, rockets, planets, and more may be
    interested in attending space camp. They can write for further infor-
    mation.

Earth Resources Observation Systems Data Center
U.S. Geological Survey
Sioux Falls, SD 57198
They have a collection of aerial photographs of the planet Earth taken from
    satellites, space shuttles, and other spacecraft. Kids can write for informa-
    tion.

NASA Jet Propulsion Laboratory
California Institute of Technology
Pasadena, CA 91109
They have information and printed materials on the travels and discover-
    ies of spacecraft such as *Voyager* and *Viking*.

Music and Movement

Play some of the music or popular songs that deal with the theme of space,
such as "Aquarius" from the musical *Hair,* or the themes from *Star Wars, Star
Trek,* or *2001: A Space Odyssey.* Discuss how the music captures the mood of
space. Encourage youngsters to either create their own modern dance, inspired
by one of the pieces of music, or capture the mood through some type of
artistic endeavor.

✪ **In addition to our satellites, there are about 5 million tons of space junk circling the Earth.**
  - *How many pounds of space junk circle the Earth?*
  - *What does most of that space junk consist of?*
  - *What will happen to most of that space junk?*

✪ **Every day, a total of approximately 15,000 tons of meteors fall to the Earth. Most are smaller than a grain of rice.**
  - *Where do all those meteors come from?*
  - *What happens to most of the meteors before they land on Earth?*
  - *How many meteors fall to Earth every year?*

✪ **Earth is the only planet with a single moon.**
  - *How many planets have multiple moons?*
  - *How many planets have no moons?*
  - *Which planet has the most moons?*

✪ **Canada contains $1/3$ of all the freshwater on Earth.**
  - *How large is Canada?*
  - *How much freshwater is contained within the United States?*
  - *Where does most freshwater come from?*

Related Literature

Barrett, Norman. *The Picture Book of Rockets and Satellites.* **(New York: Watts, 1990).**
This book is a colorful and interesting introduction to rockets and satellites.

_____ . *The Picture Book of Space Voyages.* **(New York: Watts, 1990).**
Filled with photographs and illustrations, this book offers a complete examination of space exploration.

Berliner, Don. *Our Future in Space.* **(Minneapolis, MN: Lerner, 1991).**
Lots of information on future projects and space exploration in the future.

Ficher, George. *The Space Shuttle.* **(New York: Watts, 1990).**
This book is a basic overview about the entire space shuttle program and the information collected during these voyages.

Vogt, Gregory. *Space Stations.* **(New York: Watts, 1990).**
Presents a complete overview of space stations from science fiction to science fact.

Chapter 25

Space

Carole Stott

New York: Watts, 1995

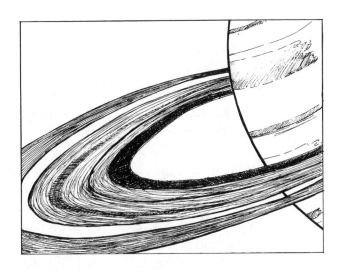

Book Summary

A thorough introduction to space science, including descriptions of the planets, the solar system, the universe, and other celestial bodies are presented in the pages of this colorful book. Crisp writing and loads of photographs and illustrations make this a perfect addition to the space science library. The book also features "see-through pages" allowing readers to interact with some of the data presented. A great "starter" book for the young scientist.

Questions to Share

1. What was the most amazing fact you learned about our solar system?
2. What else would you have liked the author to include in this book?
3. Which of the planets did you find most interesting?
4. Why is it important for us to know about the planets, even though they are so far away?

Major Book Project

Invite students to create a brief skit based on information obtained in this and other books about the planets. This skit can be based on the premise that the youngsters have been involved in a ten-year mission to a planet of their choice. As returning astronauts they are ready to tell the world of their findings. The skit (5 to 6 minutes in length) should be a mixture of fact and fiction. For example, children returning from Neptune might report on the underground homes of the "Neptunians" that were built because the strong winds and freezing temperatures on the planet make life on the surface impossible. Youngsters can also discuss some of the hardships of their visit and the technological innovations that were needed to deal with the realities of the planet or sun. It is possible that not all missions were successful. For example, children can imagine that the sun mission has to be aborted due to the fact that their specially made space suits absorbed too much heat. However, youngsters would still be responsible for sharing information about the sun and their mission there.

Special Project

Invite children to create a "Space—The Final Frontier" time line. As children read this book and others (see related literature) about the advances in space exploration, invite them to add events, such as the various manned space programs (Mercury, Gemini, Apollo, Apollo-Soyuz, Sky Lab), space shuttles (*Columbia, Challenger, Discovery, Atlantis, Endeavor*), and unmanned NASA Planetary Exploration Missions (Mariner, Magellan, Pioneer, Viking, Voyager) to the time line. Discoveries of the planets and their satellites can also be included. For each event or discovery, have children include appropriate names, dates, specifics, missions purpose (if applicable), and each mission or discovery's significance. The time line can be prepared on poster paper cut to a length of 6 feet or more. Events and discoveries can be illustrated as necessary.

Field Trip

Help youngsters appreciate the relative distances between the planets with this activity. Take a group of children to the local high school football field. Have one youngster (who will represent the "sun") stand on the end zone line at one end of the field. Ask nine other individuals to take on the roles of the nine planets and to place themselves at the following distances from the "sun." (If you don't have a sufficient number of youngsters, invite a child to lay on a piece of thick cardboard. Trace his/her body outline on the paper and cut it out. Tape two pencils along the bottom so that they can be stuck in the ground and the "body" will be able to stand upright [you may wish to make several so that the number of children plus the number of "models" equals nine].)

Planet (person)	Distance from "Sun"
Mercury	2^1/$_2$ feet
Venus	4^1/$_2$ feet
Earth	6^1/$_2$ feet
Mars	10 feet
Jupiter	11^1/$_2$ yards
Saturn	20^1/$_2$ yards
Uranus	41^1/$_2$ yards
Neptune	65 yards
Pluto	86 yards

Take time to discuss with youngsters the fact that the sun is the center of our solar system. Therefore, the planets are not strung out in a straight line as in this re-creation. If there is a very large open field available, you can set up this demonstration with the "planets" arranged in many different directions.

Write Away

Children may wish to obtain some astronomical charts and posters of the solar system and other celestial objects. The following companies can be contacted about the availability and cost of selected items. Children should write to them, obtain catalogs, and inquire about their various offerings.

Celestial Arts
231 Adrian Road
Millbrae, CA 94030

Nature Company
P.O. Box 7137
Berkeley, CA 94707

Sky Publishing Company
49 Bay State Road
Cambridge, MA 02238

Edmund Scientific
101 East Gloucester Pike
Barrington, NJ 08007

Astronomical Society of the Pacific
1290 24th Avenue
San Francisco, CA 94122

Across the Curriculum

Language Arts

Invite youngsters to gather a collection of related books on the solar system (see related literature). Encourage them to assemble a collection of incredible facts about the planets in our solar system. These facts can be set up in a manner similar to that included in this book or can be prepared as a large wall chart to be added to over a long period of time.

For each planet, invite youngsters to include one or more incredible facts. If possible, obtain or prepare beforehand a large illustrated wall chart of the major celestial bodies. The factual data can be printed on small strips of paper and attached to the chart with pins or tape. Encourage the addition of new facts as more information is learned about the planets. The following data (obtained from *Space*) will help youngsters get started:

Sun—The sun is so big that 110 Earths would fit across its face.
Mercury—A year on Mercury lasts only 88 days.
Venus—Venus is one of only two planets that has no moons.
Earth—Earth is the only planet with water and life (that we know of).
Mars—A volcano on Mars is three times the height of Mount Everest.
Jupiter—A day on Jupiter lasts only ten hours.
Saturn—Winds up to 1,120 m.p.h. have been recorded on Saturn.
Uranus—This planet is the only one that spins on its side.
Neptune—Neptune has a moon that orbits the planet every seventeen hours.
Pluto—Pluto is the farthest planet from the sun.

✪ **All things considered, 99 percent of the entire universe is actually nothing.**
- *Besides planets, what other celestial bodies fly through space?*
- *How big is the universe?*
- *What is the largest object in space?*

✪ **The Earth receives only ¹/₂ of one-billionth of the sun's radiant energy.**
- *What happens to the rest of the sun's energy?*
- *Are there other celestial bodies that produce energy?*
- *How long does it take sunlight to reach the Earth?*

✪ **Jupiter is more than 2¹/₂ times larger than all the other planets of our solar system combined.**
- *How much larger is Jupiter than Earth?*
- *What is the smallest planet in our solar system?*
- *Do you believe life could exist on Jupiter?*

✪ **The original name for the planet Uranus was Herschel.**
- *How did the planet Mars get its name?*
- *What is another name for the planet Venus?*
- *If a new planet is discovered, what should its name be?*

Related Literature

Asimov, Issac. *How Did We Find Out about Neptune?* **(New York: Walker, 1990).**
This book describes the marvelous way this planet was discovered, with a focus on observation and prediction.

Gallant, Roy. *The Macmillan Book of Astronomy.* **(New York: Macmillan, 1986).**
A complete guide to all the planets, stars, asteroids, comets, and meteors of our solar system.

Kelch, Joseph. *Small Worlds: Exploring the 60 Moons of Our Solar System.* **(New York: Messner, 1990).**
An inviting look at the sixty moons spread throughout our solar system.

Lampton, Christopher. *Stars and Planets.* **(New York: Doubleday, 1988).**
Everything a young reader would want to know about the solar system is in this complete book.

Mitton, Jacqueline. *Discovering the Planets.* **(Mahwah, NJ: Troll, 1991).**
Good explanations and lots of information about the planets is included here along with clear illustrations.

Simon, Seymour. *Galaxies.* **(New York: Morrow, 1988).**
The formation and location of galaxies is highlighted by marvelous photographs.

———
 . *Mars.* **(New York: Morrow, 1987).**
The "red planet" is explored in wonderful detail and colorful photographs. One of a series.

Chapter 26

The Sun

Seymour Simon

New York: Mulberry, 1986

Book Summary

Noted science author Seymour Simon presents an up-close and personal look at the largest star in our solar system. He examines the wonders of the sun, its constant nuclear explosions, and its sea of boiling gases that form the surface. A crisp, clear text highlighted with more than twenty startling, full-color photographs, this is a fascinating introduction to this magnificent star.

Questions to Share

1. What was the most amazing fact you discovered about the sun?
2. Imagine that the sun no longer existed. How would life on Earth be different?
3. What did you learn about the sun from the color photographs?
4. Why do people find a solar eclipse so fascinating?
5. Do you believe that there are other stars like the sun in the universe? What information do you have to back up your thoughts?

Major Book Project

The following sun-related activities will help youngsters learn more about this magnificent star:

Star Path

Invite youngsters to place a sheet of paper outside (on a sunny day) on a table top. Ask them to make an "X" in the center of the paper. Have them place an upside-down glass bowl (2-quart size) on top of the paper with the "X" in the center of the bowl. Invite them to touch the glass dome with the tip of a pencil so that the shadow of the pencil's tip falls on the "X" mark. Using a marking pen, invite children to make a mark on the glass at the precise point where the tip of the pencil touches the glass. Have children continue to make marks on the glass throughout the day at one hour intervals.

(Children will note a "path" of dots across the underside of the bowl. It seems as though the sun is moving across the sky in an east to west path. However, the sun is not moving, rather it is the Earth that is turning toward the east. Since the Earth has a full rotation every 24 hours, it gives the illusion that the sun rises in the east and sets in the west. In fact, the sun is stationary—the rotation of the Earth makes it seem as though the sun moves and we are stationary.)

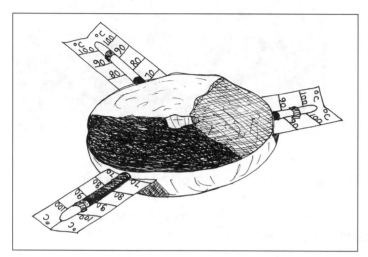

Bagel-o-meter

Invite children to cut a large bagel in half lengthwise. Divide the surface of one bagel into three sections. Using three different colors of tempera paint (i.e., black, orange, white [a dark color, a medium color, a light color]) have youngsters paint each of the sections a different color. Obtain three inexpensive thermometers (from a teacher supply store, a variety store, or a drug store) and stick one thermometer in each of the three sections (see the accompanying illustration). Place this "bagel-o-meter" outside at the beginning of a sunny day. Invite children to record the temperatures on each of the three thermometers throughout the day at one hour intervals. Invite them to record their observations in a journal or notebook. Plan to discuss any differences between the three thermometers.

(Dark objects absorb more heat [sunlight] than light objects. The thermometer stuck into the section of bagel that was painted black should consistently indicate a higher temperature reading than the thermometer stuck into the bagel section painted white. This is why people tend to wear more dark-colored clothing in the winter and more light-colored clothing in the summer.)

Eclipse Me

Children can set up some conditions that simulate a solar eclipse. Invite them to stand outside and locate a large object (a house, a tree, a building) in the distance. Have them put a hand over one eye and hold a quarter at arm's length in front of their other eye (while looking at the object). Instruct them to bring the coin closer and closer to their face until it is directly in front of the open eye. They will note that as the quarter approaches their face, less of the object is seen until eventually the entire object is "covered" by the coin.

(Although the quarter is smaller than the object, it is able to "cover" that object when it is close to the observer. Although the moon is smaller than the sun, it is able to "block out" the sun because it is closer to us and in a direct line between the Earth and the sun. This is called a solar eclipse, but it doesn't occur regularly because the moon's orbit is not around the equator [thus putting it in a direct line] and the Earth's axis is tilted at $23\frac{1}{2}°$.)

(*Note:* Here are some dates of upcoming solar eclipses and where they will occur: February 26, 1998—Panama, Columbia, Venezuela, Guadeloupe,

Montserrat, Antigua; August 11, 1999—Europe, the Middle East, India; June 21, 2001—the Atlantic Ocean, Southern Africa; December 4, 2002—Southern Africa, the Indian Ocean, Australia.)

Arts and Crafts

Invite youngsters to create a series of sun pictures. These can be done quite easily and can result in a number of innovative and creative products.

Obtain several sheets of 12- x 18-inch colored construction paper from an arts and crafts store or a variety store. Using a pattern (or a manufactured stencil—also available at a crafts store) invite children to cut out the letters of their names from a blank sheet of paper. Have them place the letters on top of the large sheet of construction paper and lay the construction paper outside on a sunny day.

At the end of the day, have children retrieve their paper and bring them inside. When they remove the letters from the construction paper they will note that the area around the letters faded as a result of being in the sunlight. Children will be able to easily see the outline of the letters they used. These "pictures" can be displayed for others to enjoy.

Children may also wish to create other types of sun pictures using stencils of animals or similar objects.

Across the Curriculum

Language Arts

Invite children to create one or more sets of pyramid sentences using information from the book. Following are two examples of pyramid sentence sets:

Sun
Sun shines.
Summer sun sears.
Serious scorpions saunter slowly.
Sweltering sun sends surfers scurrying.

Pluto
Pluto persists.
Planets pose possibilities.
Puny Pluto passes planets.
Planet pebbles produce pleasant perusal.

The design is to begin with a one-word topic. The next line is a two-word sentence in which each word begins with the same letter as the word in the first line. The third line is a three-word sentence beginning with the identical letter, and so on. These "poems" can be continued for as long as children wish, although it is suggested that a five-line pyramid sentence set is usually sufficient.

✺ **The sun is 100,000 times brighter than the full moon.**
  • *What is a full moon?*
  • *How often does a full moon occur?*
  • *How much larger is the sun than the moon?*

✺ **The earliest recorded solar eclipse occurred on October 22, 2137 B.C.**
  • *How many solar eclipses have occurred in the 20th century?*
  • *What was the most recent solar eclipse in the United States?*
  • *What must happen in order for a solar eclipse to occur?*

✺ **It takes the sun about 230 million years to orbit once around the center of the Milky Way galaxy.**
  • *How long does it take the Earth to orbit the sun?*
  • *How long does it take the moon to orbit the Earth?*
  • *What planet takes the longest time to orbit the sun?*

✺ **Every second, on each square yard of its surface, the sun produces enough energy to light 100,000 homes.**
  • *How much energy does it take to light your home?*
  • *How much does that energy cost?*
  • *How much would you save if you converted sunlight to solar energy?*

Related Literature

Barrett, Norman. *The Picture World of Sun and Stars.* (New York: Watts, 1990).
Straightforward information about the sun as a star, its temperature, size, composition, and place in the Milky Way.

Bendick, Jeanne. *The Sun, Our Very Own Star.* (Brookfield, MA: Millbrook Press, 1991).
The location of the sun in the universe, as well as its power and size, are explained in simple terms.

Burns, George. *Exploring the World of Astronomy.* (New York: Watts, 1995).
Lots to discover and a wide range of activities highlight this all-inclusive book.

George, Michael. *The Sun.* (Mankato, MN: Creative Education, 1991).
A beautiful book that relates facts and information in an artistic, narrative form.

Gustafson, John. *Planets, Moons, and Meteors.* (New York: Messner, 1992).
A detailed and descriptive look at the planets and other celestial bodies in our solar system.

Stott, Carole. *Space.* (New York: Watts, 1995).
A richly illustrated (with see-through pages) and informative text about the solar system (see chapter 25).

Part IX

General—
This and That

Chapter 27

Brain Surgery for Beginners

Steve Parker

BROOKFIELD, CT: MILLBROOK PRESS, 1993

Book Summary

The human brain controls every part of the body from head to toe. It moves muscles, moves blood, and moves air in and out of the body. It helps the body fill up and empty out. It recognizes sights and sounds and converts them into recognizable images. This light-hearted book, filled with amusing and descriptive illustrations, looks at all aspects of the human body from the vantage point of the brain. Facts and figures are cleverly presented and designed to help kids learn all about the most important and complex machine in the world.

Questions to Share

1. Why did the author give this book the title he did?
2. What was the most amazing thing you discovered about the human body?
3. Which part or operation of your body would you like to learn more about?
4. Which features of your body make you similar to your friends? Which features are distinctly different?
5. Besides doctors, who are some other people who help you take care of your body (or parts of your body)?

Major Book Project

The human brain is, arguably, the most important body organ there is. As children read this book they will learn about this magnificent organ, what it does, and what it controls. This "light" approach to anatomy and physiology will impress youngsters with the magnificence of this body part.

Few youngsters have ever seen a brain. You can provide them with an opportunity to do so, and, thus, help them understand some of the complexities and intricacies of this organ. Visit a local butcher shop and obtain a calf's brain (the shape of a calf's brain is similar to that of a human's). Lay sheets of newsprint on top of a table and place the calf's brain in the middle. Using a sharp kitchen knife, cut into the brain (you may wish to do the cutting while children refer to the illustrations of various brain parts in *Brain Surgery for Beginners*).

As you are doing the cutting, take time to discuss the similarities between the calf's brain and the human brain. What external or internal features are identical? Does a calf's brain perform the same functions as a human brain? What is a human brain able to do that a calf's brain cannot?

During the cutting process, youngsters may wish to take a series of photographs. Later, these photos can be mounted in a scrapbook along with accompanying comments or descriptions. Children may wish to refer to *Brain Surgery for Beginners* for labels and relevant information on selected features. Additional comparisons can be made between the calf's brain and photographs of the human brain found in encyclopedias.

Youngsters may wish to complete charts similar to the ones below:

Human Brain	
Features	What It Does/Controls
Calf Brain	
Features	What It Does/Controls

Field Trips

Select one or more of the following field trips to share with children. Depending on the size and location of your local community, additional visits to other sites (with accompanying interviews) may also be possible. Check with your local health organization, hospital, doctor referral service, or visiting nurses association. They will be glad to refer you to additional groups, organizations, and specialists who are pleased to answer youngster's questions and provide relevant tours or demonstrations.

A. If possible, obtain permission to visit a local rehabilitation center or chiropractor. Encourage children to talk with the personnel about the structure and function of the human skeleton and muscle system. What are some of the ways with which humans can protect their bones and muscles? What are some possible exercises and dietary habits? What happens to people who suffer from diseases or injuries to their bones or muscles? The data that children collect can be assembled into a scrapbook or notebook.

B. Take children to visit a family doctor or health clinic. Invite them to obtain information on the suggested heights and weights of children at different ages. Help them understand that these suggested measurements are only averages, and might be different from the height and weight of the youngsters with whom you work. Encourage children to talk with a physician about the diet, sleep, and exercise children need to maintain proper growth and development.

C. If possible, visit a nearby blood bank and encourage youngsters to talk with one of the technicians or nurses. Invite them to find out how blood is collected, measured, stored, and preserved. What precautions do the workers have to follow? How much blood is collected in a day, a week, or a month? How is that blood used? How is it transported? In light of the AIDS epidemic, how is the blood tested and protected?

Write Away

Invite youngsters to write to one or more of the following organizations and request some relevant literature on the human body and its care. Plan time to discuss the information with children and encourage them to assemble the data into a notebook or scrapbook.

American Speech-Language-Hearing Association
10801 Rockville Pike
Rockville, MD 20852
They have some descriptive literature on the care of eyes and ears.

National Institutes of Health
Building 1, Room 2B19
Bethesda, MD 20892
Request information and brochures on the diseases of the lungs and kidneys.

National Clearinghouse for Alcohol and Drug Information
P.O. Box 2345
Rockville, MD 20852
They have a free catalog of materials available to share with children.

National AIDS Information Clearinghouse
P.O. Box 6003
Rockville, MD 20849
Write and obtain the "AIDS Prevention Guide," which has ideas on how to talk about this disease and accurate answers to common questions.

National Institute of Child Health and Human Development
National Institutes of Health
Building 31, Room 2A32
9000 Rockville Pike
Bethesda, MD 20892
Write and obtain a listing of the brochures and information sheets on mental health for children.

Across the Curriculum

Language Arts

Invite children to select one or more of the following topics and discuss their creative insights. They may wish to write a brief report on their personal interpretation. Obviously, there are no right or wrong answers to these "investigations"; however, youngsters should feel free to use the data and information in *Brain Surgery for Beginners* to arrive at some conclusions or suppositions.

1. Discuss or write a story from the viewpoint of a body organ, such as the heart, lungs, or stomach.
2. Prepare a time line or storyboard on "A Day in the Life of My (body organ)."
3. "My favorite body part"
4. If I could look inside my body I would like to see …
5. "The care and feeding of the human brain"

Experiments

Children may be interested in selecting one or more of the following experiments to learn more about their own bodies.

A. Roll some clay into a ball about the size of a marble. Stick a wooden match into the ball. Invite youngsters to place this device on their wrist (match sticking up). They may need to move it around until they find a spot with a strong beat. Encourage them to record the number of beats in 30 seconds and multiply by 2 to obtain their heartbeat. Invite several youngsters to do this and record their heartbeats. How do you account for any differences between children? Encourage children to participate in selected physical activities for 2 to 3 minutes (stepping up and down on a stair, jumping rope, doing continuous jumping jacks) and record their heartbeats 1 minute after, 3 minutes after, and 5 minutes after. The data can be recorded on separate charts for each individual.

B. Inform children that the human tongue has four different types of taste buds. These include sour buds, bitter buds, salty buds, and sweet buds. Each set of taste buds is located on a different portion of the tongue. Children can map out portions of their tongues as follows. Obtain several clean cotton swabs. Invite youngsters to dip each of four cotton swabs into the following solutions and then touch those tips to various portions of their tongues.

- lemon juice (sour)
- tonic water (bitter)
- salt water (salty)
- sugar water or corn syrup (sweet)

Youngsters may wish to draw an oversize illustration of a human tongue and plot the location of the four major types of taste buds.

C. Mix together ¹/₂ cup of water and 2 teaspoons of corn starch. Stir well. Cut some paper towels into several 2- x 2-inch squares and dip them in the liquid. Set them aside to dry. Work with youngsters and paint one of their palms with iodine (this should be done only by an adult). Encourage youngsters to engage in a physical activity for a period of time (they should build up a sweat). Invite them to place one of the paper towel squares on their iodine-covered palm. They will notice the sweat glands on that palm showing up as dark spots and that sweat glands seem to be concentrated in selected areas of the skin.

D. In each of three small plastic cups pour 3 tablespoons of milk. In the first cup put 2 tablespoons of water. Cover the cup with a sheet of plastic wrap, using a rubber band to hold the wrap in place. In the second cup, put 2 tablespoons of a weak acid such as lemon juice or vinegar and cover as above. In the third cup put 2 tablespoons of an enzyme such as a meat tenderizer and cover as above. After one or two hours invite youngsters to observe the changes that occur in each cup. The changes that occurred in cups two and three are similar to the digestive process in the human stomach.

⚡ **In one year, the human heart will beat about 36 million times.**
- *How many times does your heart beat in one day?*
- *What causes a heart to beat faster than normal?*
- *What can slow down a human heart beat?*

⚡ **A nerve impulse travels to the human brain at a speed of 205 miles per hour.**
- *What is the fastest you have ever traveled (in a car, boat, airplane)?*
- *Why is it necessary for a nerve impulse to travel so fast?*
- *What might happen if the impulse was not as fast?*

⚡ **The average person takes approximately 23,040 breaths every day.**
- *How many breaths do you take in a minute? In an hour?*
- *How is your breathing related to your heart rate?*
- *What animal breaths the least?*

⚡ **Forty-three percent of the human body is muscle.**
- *How much of your body is bone?*
- *How much of your body is water?*
- *How much of your body is skin?*

Related Literature

Allison, Linda. *Blood and Guts: A Working Guide to Your Own Insides.* (Boston, MA: Little, Brown, 1976).
This book remains one of the finest and most complete introductions to the human body as seen and understood by kids.

Bruun, Ruth, and Bertel Bruun. *The Brain: What It Is, What It Does.* (New York: Greenwillow, 1989).
Packed with lots of information about the human brain, this book provides much insight into this marvelous machine.

Markle, Sandra. *Outside and Inside You.* (New York: Bradbury, 1991).
The human body is seen through a collection of X-rays, electron microscopes, thermograms, and other images.

Parker, Steve. *The Body Atlas.* (New York: Dorling Kindersley, 1993).
Body layers are "peeled" away in marvelous illustrations accompanied by conversational insights into the features and functions of the human body.

Settel, Joanne, and Nancy Baggett. *Why Does My Nose Run? and Other Questions Kids Ask about Their Bodies.* (New York: Atheneum, 1985).
All the answers to questions kids have about their bodies can be found in this delightful book.

Chapter 28

The Case of the Mummified Pigs and Other Mysteries in Nature

Susan Quinlan

HONESDALE, PA:
BOYDS MILLS PRESS, 1995

Book Summary

Why would a herd of reindeer suddenly die off? What would cause a change in the color of a wild songbird's tail? What would cause dead pigs to turn into mummies? These are just a few of the mysteries investigated and solved in this intriguing book. Readers follow the steps of scientists from around the world as they track down clues and discover some of the amazing ways that nature works. This book is an incredible insight into "science in action"—the mysteries of nature and how scientists work together to solve those mysteries every day.

Questions to Share

1. Which of the mysteries presented in this book did you enjoy the most?
2. What are some of the procedures or processes that scientists use to solve nature mysteries?
3. Which mystery do you believe was the most difficult for scientists to solve?
4. If you were given the opportunity, which mystery would you have enjoyed being a part of?
5. What are some nature mysteries in and around where you live that you would like to have solved?

Major Book Project

Invite youngsters to select one or more of the following activities, each of which illustrates a principle or process presented in *The Case of the Mummified*

Pigs. After children have had an opportunity to do these projects, invite them to create additional activities for other mysteries presented in this book.

A. (*The Case of the Mummified Pigs*) Invite youngsters to take two slices of bread and wet them slightly (don't soak them). Carefully rub one slice across a kitchen table or countertop (do this carefully so that the bread does not tear). Place the slice of bread in a sealable plastic sandwich bag and seal it. Take the second moistened slice of bread and gently rub it over the surface of the kitchen floor, place it in a bag, and seal it. Have youngsters place a third, completely dry, slice of bread in another sandwich bag and seal it. Invite kids to put all three bags in a warm and dark location (a closet) for a couple of days.

(Molds and other microscopic plants are everywhere [even in the cleanest kitchen!]. Two of the bread slices will begin to decompose because the conditions for mold growth—warmth, darkness, moisture—are available. This natural process takes place all the time in nature.)

B. (*The Mice, the Ants, and the Desert Plants*) Invite youngsters to locate a piece of rotting wood or an old log outside. Have them collect lots of ants by scooping them into a large glass jar. Include some of the nearby soil. Wrap the jar with black construction paper so that it is completely dark inside. Put some water in a large cake pan and put a saucer upside down in the middle of the pan. Place the jar on the upturned saucer (the water prevents the ants from escaping). Invite youngsters to sprinkle some sugar water over the soil, place two or three small bits of fruit inside, and place the lid on the jar.

(The ants will begin to dig their tunnels in the soil. If it is dark enough they will construct those tunnels around the sides of the jar. If the black construction paper is removed every week or so, youngsters will be able to see the progress the ants are making in constructing their tunnels.)

C. (*The Mystery of the Missing Songs*) Obtain a piece of lumber that is 2 inches square on each side and approximately 14 to 16 inches long. Using a 1-inch drill bit, drill a series of holes down each side (each hole should be approximately 1 inch deep). Fasten an eye screw to one end of the stick and pass a long piece of strong string through it. Stuff the holes of this bird feeder with a sticky bird food (see recipe below) and hang the feeder in a nearby tree. Invite youngsters to observe the number and variety of birds that visit the feeder over a predetermined length of time (one week, one month).

To make bird food, mix one cup of suet (from the butcher shop), one cup of chopped nuts, one cup of sunflower seeds, one cup of chunky peanut butter, one cup of cornmeal, and one tablespoon of crushed eggshells together. Stir thoroughly and refrigerate in a plastic container. Spoon some of the mixture into the holes in the feeder constructed above.

Special Project

One of the ways scientists are able to solve mysteries about animals is by examining the tracks they leave behind. Often, these tracks are left in soft mud or clay. By closely observing these prints, scientists are able to determine the height and weight of various animals as well as other pertinent information.

This activity can be done shortly after it has rained or whenever you might find some animal tracks. Youngsters will need a large bowl, water, a strip of cardboard that can be formed into a ring, a small ruler or tape measure, tape, a box of plaster of paris (which can be obtained at any hardware store), paper, and pencil.

1. Have children prepare rings by joining the ends of several cardboard strips together with tape. A good size for the rings would be about 5 to 6 inches in diameter.
2. Gather the other supplies together and take youngsters outside to look for any animal prints that may be on the ground. If you live in a city, you may discover prints of dogs or cats. If you live in the suburbs or the country, prints of other animals may be available.
3. Share some observations about any prints discovered.
   "What type of animal do you think made this print?"
   "Why do you think that?"
   "What do you notice that is special about this print?"
   "How is it similar or different from your pet's prints?"
4. Have children carefully measure the prints (length, width, and depth) and record that information on a sheet of paper.
   "What do we know about the environmental conditions when this animal walked here?"
   "What else can we learn about this animal?"
5. Ask youngsters to carefully place their cardboard rings around the prints and press each ring into the ground. Mix the plaster of paris according to the package directions.
   "Why is it important for us to follow the directions on the package?"
   "What might happen if we didn't follow those directions?"
6. Pour the plaster of paris mixture into each ring and let it dry for about 30 minutes.
   "Why should the mixture dry before we lift it up?"
7. Pick up each cast and place it carefully in a cardboard box. If you wish, children may want to make a cast of other nearby prints that are different from their first ones.
8. Later, brush off the dirt and remove the ring from each cast. Examine the impressions with children. Record any information on sheets of paper.
   "What do you notice about this print?"
   "Are there some things you see now that you didn't see when we looked at the original print?"

Scientists take plaster casts of animal prints in order to study them in the laboratory. The casts are an accurate record of an animal's footprint and yields valuable data for the scientist. Children can help solve mysteries related to animals that walk through their communities or neighborhoods by "collecting" and closely examining those tracks.

Write Away

Invite youngsters to write a letter to the Office of Endangered Species, Fish and Wildlife Services, U.S. Department of the Interior, Washington, D.C. 20241. Encourage children to ask for a copy of the list of endangered animal species. When the list is received, invite youngsters to check to see which, if any, of the species on the list are in their area of the country. Children may wish to develop an "action plan" for helping to protect the species living in their area. The plan may take the form of a campaign, posters, writing letters, and so on. If none of the endangered species live in the local area, invite children to select a species that lives in an area close by to be the focus of the activity.

⚡ **There are more than twelve thousand different varieties of ants in the world.**
- *Why do you think there are so many different types of ants in the world?*
- *What is the total number of animal species in the world?*
- *What is the total number of plant species in the world?*

⚡ **More than five hundred species of plants subsist partly on decomposed animal tissue.**
- *What are the benefits to the environment from these plants?*
- *What else causes animals to decompose?*

⚡ **The typical plant or tree receives approximately 10 percent of its nutrition from the soil. The rest comes from the atmosphere.**
- *What are some basic nutrients all plants need?*
- *What happens to plants when one or more nutrients are missing?*
- *How do humans assist plants in getting their proper nutrients?*

⚡ **Peregrine falcons can reach speeds of 180 miles per hour.**
- *What are some of the fastest birds in the world?*
- *What are the advantages of flying fast?*

Related Literature

Cook, David. *Environment*. (New York: Crown, 1985).
How can we conserve our natural resources and preserve plant and animal life?

Florian, Douglas. *Nature Walk*. (New York: Greenwillow, 1989).
The things that one can discover in a simple walk through the forest are featured in this enlightening book.

Herberman, Ethan. *The City Kid's Field Guide*. (New York: Simon & Schuster, 1989).
A wonderfully insightful guide to the flora and fauna of an urban environment.

Pringle, Lawrence. *Saving Our Wildlife*. (Hillside, NJ: Enslow, 1990).
North American wildlife and our efforts at preserving many species are the emphases in this engrossing book.

O'Neill, Mary. *Nature in Danger*. (Mahwah, NJ: Troll, 1991).
This book describes how ecosystems function and the role humans play in those processes.

Chapter 29

53¹/₂ Things That Changed the World and Some That Didn't

Steve Parker

BROOKFIELD, CT:
MILLBROOK, 1992

Book Summary

This book could be appropriately subtitled "Science in Action" or "Science in Our Everyday Lives." From common inventions such as the toilet, the clock, the screw, the telephone, the airplane, and the camera to less familiar (but always useful) inventions such as the blast furnace, the combine harvester, and fusion power, this book provides readers with some incredible insights into the devices and objects that have made our lives easier and more comfortable. Each invention is described in colorful detail on one or two oversize pages with the emphasis of history, associated problems, how it works, and its effects on modern life. A great resource for the science library.

Questions to Share

1. Which of the inventions did you find to be the most amazing?
2. What invention should have been included in this book (but wasn't)?
3. Which of the inventions included in this book do you use most often on a daily basis?
4. If you could meet any one of the inventors detailed in this book, which one would it be? What would you want to discuss with that person?
5. Why were most of the early inventors men?

Major Book Project

Invite youngsters to participate in an "invention convention." Encourage them to select one or more of the following activities and record their observations, experiments, and results in a journal or notebook. Be sure to take time to discuss their reactions to any successes as well as their feelings about any failures. It will be important for children to understand that scientists experience a lot of failures before they successfully solve a problem (Thomas Edison failed 1,600 times before he finally invented the incandescent lightbulb—something we all use and often take for granted every day). Let children know that scientists come up with lots of ideas before discovering the single idea that will solve their problem.

A. Provide youngsters with a simple, everyday object and ask them to create ten new uses for it. For example, a wire coat hanger can become a giant safety pin, a back scratcher, a bath towel holder, a giant cookie cutter, a very large toothpick, a baton, a plant hanger, a free-form sculpture, a wand, or a paper picker-upper.

B. Invite children to select one of the inventions detailed in the book and prepare a brief oral report on the significance or effects that item has had on society. For example, how has the telephone changed the way people communicate with each other? How has it made our lives easier? In what ways has it allowed us to do things we previously were unable to do?

C. Invite youngsters to select one of the inventions mentioned in the book and to create a time line of the significant events in the discovery, development, and use of that invention. Kids can draw a long horizontal line on a strip of newsprint or poster board. Major events in the history of the selected invention can be recorded on the line in chronological order from left to right. Youngsters may wish to add other historical or social events that occurred in and around the listed dates on the time line. These can provide important points of reference for the invention's development.

D. Many of the inventions we use quite frequently today have been heavily advertised. However, inventions made many years ago did not benefit from active advertising campaigns. Invite children to select one of the earlier inventions and develop an appropriate advertising campaign to "sell" the invention to the public. For example, what types of advertising or promotion would have made people in the 1600s want to purchase a microscope? Encourage children to look at magazine advertisements or TV commercials for examples of advertising "gimmicks" that help promote and sell a product.

E. Invite students to select any five of the inventions mentioned in the book and explain how they would be able to live or survive today if that invention had not been created many years ago. For example, how would their lives be different if the plow had never been invented? If nobody had ever designed the screw, how would our lives be different today? Youngsters may wish to present an oral report or a brief one-page description of their ideas.

F. Provide youngsters with a box of old machine parts, parts of a broken appliance or clock, levers, gears, cranks, or any potpourri of mechanical parts obtained at a yard or garage sale or neighborhood junk shop. Invite students to create their own original invention (a "homework machine" for example). What will the invention do? Who will be able to use it? How will it work? How long will it last?

As youngsters participate in the events of the "invention convention" plan time for them to discuss their observations and viewpoints. It might be appropriate to discuss some of the qualities of a good inventor. What personality characteristics do good inventors share? Can anybody be an inventor?

Special Projects

A. Youngsters may wish to create their own "car" with the following activity. Push one end of a rubber band through the hole in a spool of thread (this can be accomplished by pushing it through the hole using a large paper clip that has been straightened out). When one end of the rubber band pokes out of the other end of the spool, slip it onto a medium-sized paper clip. Tape this clip to the far end of the spool.

Slip the rubber band at the near end of the spool through a jumbo paper clip. Turn this clip with your finger to wind it. Be careful that the rubber band doesn't bunch up inside the spool.

After the rubber band is twisted into the hole, keep winding for about 15 to 25 additional turns. Place the "car" on a table top or other flat surface and release it. It will dash across the surface. A rubber band wrapped around the middle of the spool will provide more traction for the "car."

After children have created and practiced with a "car" invite them to make design changes that will make the "car" go faster, more in a straight line, or smoother. What items can they add to the car that will improve its performance? How is their invention similar to any of the inventions mentioned in the book? What are some of the practical uses of their invention? Can their invention be used as part of a larger invention? If so, what would that invention be?

B. Provide children with a list of the six basic simple machines: lever, pulley, wheel and axle, inclined plane, wedge, and screw. Invite them to look for examples of those simple machines in everyday life and to provide examples of their use. A chart similar to be one below can be prepared and given to youngsters (a few items have already been filled in):

What Is It	Examples
Lever	hammer; wheelbarrow
Pulley	
Wheel and Axle	doorknob
Inclined Plane	staircase
Wedge	
Screw	screw

Invite youngsters to consult related literature in the school or public library for examples of simple machines and how they are used in our everyday lives.

⚡ **Scientists at the University of Utah have invented a working motor that is the width of seven human hairs.**
- *Where would such a small motor be used?*
- *What advantages would a small motor have over a large motor?*
- *What is the world's largest motor?*

⚡ **A power plant in California generates 14 megawatts of electricity per hour by burning old tires.**
- *How else is electrical power generated?*
- *Which country uses the most electrical power?*
- *How can electrical power be conserved?*

⚡ **The safety pin was invented in Egypt more than 4,500 years ago.**
- *What are some other inventions made by the ancient Egyptians?*
- *Are there other "old" inventions that we still use today?*
- *What is the simplest invention you know of?*

⚡ **The first "computer"—the abacus—was invented around A.D. 300.**
- *How many computers are in use in the United States today?*
- *What are some of the everyday tasks computers do?*
- *What do you think the computer of tomorrow will be able to do?*

Related Literature

Berder, Lionel. *Invention*. (New York: Knopf, 1991).
This book deals with inventions—what they do, how they work, and who invented them. The inventions range from those created by early Egyptians to those of the present day.

Caney, Stephen. *The Invention Book*. (New York: Workman, 1985).
This book explains everyday inventions such as tea bags, pencils, straws, and so on. It also explains the inventive process from idea to marketing.

Gardner, Robert. *Experimenting with Inventions*. (New York: Watts, 1990).
Stories about inventors and the inventive process encompass the first part of this book. The second part presents activities that allow readers to become involved in their own inventions.

Parker, Steve. *Everyday Things and How They Work*. (New York: Random House, 1991).
Lots of well-documented descriptions of how things work and how they're used in our everyday lives.

Taylor, Barbara. *Machines and Movement.* **(New York: Warwick Press, 1990).**
A collection of experiments with complete explanations of the principles of mechanics highlight this book.

Wulffson, Donald. *The Invention of Ordinary Things.* **(New York: Lothrop, 1981).**
A collection of stories that trace the origin of twenty-eight common inventions from shopping carts to rubber bands.

Appendix— Science Resources

Science Periodicals for Children

The following list contains the names of some of the more popular science magazines for children. Make sure your school library subscribes to several of these. As a parent, you may want to enter a subscription to one or more of these periodicals for your child(ren).

3-2-1 Contact
Children's Television Workshop
P.O. Box 2933
Boulder, CO 80322
(10 per year)

Audubon Adventure
National Audubon Society
613 Riversville Road
Greenwich, CT 06830
(6 per year)

Chickadee
Young Naturalist Foundation
P.O. Box 11314
Des Moines, IA 50340
(10 per year)

The Curious Naturalist
Massachusetts Audubon Society
Lincoln, MA 01773
(4 per year)

Dolphin Log
Cousteau Society
8430 Santa Monica Boulevard
Los Angeles, CA 90069
(4 per year)

Electric Company
Children's Television Workshop
One Lincoln Plaza
New York, NY 10023
(10 per year)

Exploratorium Magazine
3601 Lyon Street
San Francisco, CA 94123
(4 per year)

Faces
Cobblestone Publishing, Inc.
20 Grove Street
Peterborough, NH 03458
(10 per year)

Junior Astronomer
Benjamin Adelman
4211 Colie Drive
Silver Springs, MD 20906
(6 per year)

Junior Natural History
American Museum of Natural History
New York, NY 10024
(Monthly)

Kind News
The Humane Society of the United States
2100 L Street NW
Washington, D.C. 20037
(5 per year)

My Weekly Reader
American Education Publications
Education Center
Columbus, OH 43216
(Weekly during the school year)

National Geographic World
National Geographic Society
17th and M Streets NW
Washington, D.C. 20036
(Monthly)

Odyssey
Kalmbach Publishing Company
1027 North Seventh Street
Milwaukee, WI 53233
(Monthly)

Owl
Young Naturalist Foundation
P.O. Box 11314
Des Moines, IA 50304
(10 per year)

Ranger Rick
National Wildlife Federation
8925 Leesburg Pike
Vienna, VA 22184
(Monthly)

Science Activities
4000 Albemarle Street NW
Washington, D.C. 20016
(4 per year)

Scienceland
Scienceland, Inc.
501 Fifth Avenue
New York, NY 10017
(8 per year)

Science News
Science Service, Inc.
1719 N Street NW
Washington, D.C. 20036
(Weekly)

Science Weekly
P.O. Box 70154
Washington, D.C. 20088
(18 per year)

Science World
Scholastic Magazines, Inc.
50 West 44th Street
New York, NY 10036
(Monthly)

Space Science
Benjamin Adelman
4211 Colie Drive
Silver Springs, MD 20906
(Monthly during the school year)

Wonderscience
American Chemical Society
P.O. Box 57136—West End Station
Washington, D.C. 20037
(4 per year)

Your Big Backyard
National Wildlife Federation
8925 Leesburg Pike
Vienna, VA 22184
(Monthly)

Zoobooks
Wildlife Education, Ltd.
930 West Washington Street
San Diego, CA 92103
(6 per year)

Commercial Suppliers of Science Equipment and Materials

The following businesses and science supply houses have a wealth of valuable equipment and materials for school or home use. Invite children to write to these suppliers and ask for their latest catalogs. When the catalogs arrive, plan some time to read them with youngsters and talk about materials and/or supplies that may be needed for the school or home.

Accent! Science
301 Cass Street
Saginaw, MI 48602
(517) 799-8103

Activity Resources Company, Inc.
P.O. Box 4875
Hayward, CA 94540
(415) 782-1300

AIMS Education Foundation
P.O. Box 7766
Fresno, CA 93747
(209) 291-1766

Albion Import Export, Inc.
Coolidge Bank Building
65 Main Street
Watertown, MA 02172
(617) 926-7222

American Nuclear Products, Inc.
1232 East Commercial
Springfield, MO 65803
(417) 869-4432

American Optical Instrument Division
P.O. Box 123
Buffalo, NY 14240
(716) 891-3000

American Science Center/Jerryco
601 Linden Place
Evanston, IL 60202
(773) 475-8440

Analytical Products, Inc.
P.O. Box 845
Belmont, CA 94002
(415) 592-1400

Bausch and Lomb Optical Systems Division
1400 North Goodman Street
P.O. Box 450
Rochester, NY 14692-0450
(716) 338-6005

Bel-Art Products
6 Industrial Road
Pequannock, NJ 07440
(201) 694-0500

Burt Harrison and Co.
P.O. Box 732
Weston, MA 02193
(617) 647-0647

Carolina Biological Supply Company
2700 York Road
Burlington, NC 27216
(800) 334-5551

Central Scientific Company
11222 Melrose Avenue
Franklin Park, IL 60131
(773) 451-0150

Chem Scientific, Inc.
67 Chapel Street
Newton, MA 02158
(617) 527-6626

Connecticut Valley Biological Supply Company
Valley Road
P.O. Box 326
Southampton, MA 01073
(800) 628-7748

Creative Learning Press
P.O. Box 320
Mansfield Center, CT 06350
(203) 423-8120

Creative Learning Systems, Inc.
9889 Hilbert Street, Suite E
San Diego, CA 92131
(619) 566-2880

Cuisenaire Company of America, Inc.
12 Church Street
New Rochelle, NY 10802
(800) 237-3142

Delta Education, Inc.
P.O. Box 3000
Nashua, NH 03061
(800) 258-1302

Denoyer-Geppert Science Company
5711 North Ravenswood Avenue
Chicago, IL 60646
(312) 561-9200

Edmund Scientific
101 East Gloucester Pike
Barrington, NJ 08007
(800) 222-0224

Educational Activities, Inc.
P.O. Box 392
Freeport, NY 11520
(800) 645-3739

Education Development Center
55 Chapel Street
Newton, MA 02160
(617) 969-7100

Energy Learning Center
Edison Electric Institute
1111 19th Street NW
Washington, D.C. 20036
(202) 778-6400

Energy Sciences, Inc.
16728 Oakmont Avenue
Gaithersburg, MD 20877
(301) 770-2550

Estes Industries/Hi-Flier
1295 H Street
Penrose, CO 81240
(303) 372-6565

Fisher Scientific Company
4901 West Le Moyne Street
Chicago, IL 60651
(312) 378-7770

Forestry Suppliers, Inc.
205 West Rankin Street
Jackson, MS 39204
(601) 354-3565

Frey Scientific
905 Hickory Lane
Mansfield, OH 44905
(419) 589-9905

Hubbard Scientific Company
P.O. Box 104
Northbrook, IL 60065
(800) 323-8368

Ideal School Supply Company
11000 South Lavergne Avenue
Oak Lawn, IL 60453
(312) 425-0800

Lab-Aids, Inc.
P.O. Box 158
130 Wilbur Place
Bohemia, NY 11716
(516) 567-6120

LaPine Scientific Company
6001 South Knox Avenue
Chicago, IL 60629
(312) 735-4700

Lawrence Hall of Science
University of California
Berkeley, CA 94720
(415) 642-7771

Learning Things, Inc.
68-A Broadway
P.O. Box 436
Arlington, MA 02174
(617) 646-0093

LEGO Systems, Inc.
555 Taylor Road
Enfield, CT 06082
(203) 749-2291

Let's Get Growing
General Seed and Feed Company
1900-B Commercial Way
Santa Cruz, CA 95065
(408) 476-5344

McKilligan Supply Corporation
435 Main Street
Johnson City, NY 13790
(607) 729-6511

Metrologic Instruments
143 Harding Avenue
Bellmawr, NJ 08031
(609) 933-0100

NASCO
901 Janesville Avenue
Fort Atkinson, WI 53538
(414) 563-2446

National Geographic Society
17th and M Streets NW
Washington, D.C. 20036
(202) 857-7000

National Science Teachers Association
1742 Connecticut Avenue NW
Washington, D.C. 20009
(202) 328-5800

National Wildlife Federation
8925 Leesburg Pike
Vienna, VA 22184
(800) 262-4729

Nova Scientific Corp.
P.O. Box 500
Burlington, NC 27215
(919) 229-0395

Nystrom/Eye Gate Media
3333 North Elston Avenue
Chicago, IL 463-1144
(312) 463-1144

Play-Jour, Inc.
200 Fifth Avenue, Suite 1024
New York, NY 10010
(212) 243-5200

Right Before Your Eyes
136 Ellis Hollow Creek Road
Ithica, NY 14850
(607) 277-0384

Sargent-Welsh Scientific Company
7300 North Linder Avenue
Skokie, IL 60077
(312) 677-0600

SAVI/SELPH Center for Multidisciplinary Learning
Lawrence Hall of Science
University of California
Berkeley, CA 94720
(415) 642-8941

Science Kit, Inc.
777 East Park Drive
Tonowanda, NY 14150
(716) 874-6020

The Science Man
P.O. Box 56036
Harwood Heights, IL 60656
(312) 867-4441

Soil Conservation Service
U.S. Department of Agriculture
P.O. Box 2890
Washington, D.C. 20013
(202) 217-2290

Southern Precision Instrument Company
3419 East Commerce
San Antonio, TX 78220
(512) 224-5801

The Teacher's Laboratory
214 Main Street
P.O. Box 6480
Brattleboro, VT 05301
(802) 254-3457

Tops Learning Systems
10978 South Mulino Road
Canby, OR 97013
(503) 266-8550

Turtox
5000 West 128th Place
Chicago, IL 60658
(312) 371-5500

Ward's Natural Science Establishment, Inc.
5100 West Henrietta Road
P.O. Box 92912
Rochester, NY 14692
(716) 359-2502

Wilkens-Anderson Company
4525 West Division Street
Chicago, IL 60651
(312) 384-4433

Index

D

E